SpringerBriefs in Materials

Series Editors

Sujata K. Bhatia, University of Delaware, Newark, DE, USA

Alain Diebold, Schenectady, NY, USA

Juejun Hu, Department of Materials Science and Engineering, Massachusetts Institute of Technology, Cambridge, MA, USA

Kannan M. Krishnan, University of Washington, Seattle, WA, USA

Dario Narducci, Department of Materials Science, University of Milano Bicocca, Milano, Italy

Suprakas Sinha Ray ⓘ, Centre for Nanostructures Materials, Council for Scientific and Industrial Research, Brummeria, Pretoria, South Africa

Gerhard Wilde, Altenberge, Nordrhein-Westfalen, Germany

The SpringerBriefs Series in Materials presents highly relevant, concise monographs on a wide range of topics covering fundamental advances and new applications in the field. Areas of interest include topical information on innovative, structural and functional materials and composites as well as fundamental principles, physical properties, materials theory and design.

Indexed in Scopus (2022).

SpringerBriefs present succinct summaries of cutting-edge research and practical applications across a wide spectrum of fields. Featuring compact volumes of 50 to 125 pages, the series covers a range of content from professional to academic. Typical topics might include

- A timely report of state-of-the art analytical techniques
- A bridge between new research results, as published in journal articles, and a contextual literature review
- A snapshot of a hot or emerging topic
- An in-depth case study or clinical example
- A presentation of core concepts that students must understand in order to make independent contributions

Briefs are characterized by fast, global electronic dissemination, standard publishing contracts, standardized manuscript preparation and formatting guidelines, and expedited production schedules.

Jean-Claude Tedenac

Thermodynamics of Crystalline Materials

From Nano to Macro

 Springer

Jean-Claude Tedenac
Université de Montpellier
Montpellier, France

ISSN 2192-1091 ISSN 2192-1105 (electronic)
SpringerBriefs in Materials
ISBN 978-3-030-99026-8 ISBN 978-3-030-99027-5 (eBook)
https://doi.org/10.1007/978-3-030-99027-5

This Springer imprint is published by the registered company Springer Nature Switzerland AG
The registered company address is: Gewerbestrasse 11, 6330 Cham, Switzerland

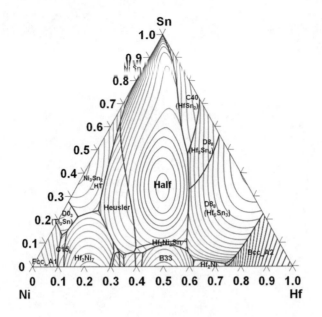

Acknowledgments

I would like to acknowledge some colleagues and friends with whom I had in-depth scientific exchanges. All of them were influencing me, each in their own way.

Firstly, Prof. Dr. Catherine Colinet from SIMAP (INP Grenoble). For a long time, we had a lot of exchanges and discussions concerning the field of Thermodynamics of Materials and Ab initio calculations.

Secondly, Dr. Pascal Yot (University of Montpellier) was a former student in my courses. He checked precisely the crystallography chapter and corrected the draft of all chapters of the document.

I would not want to forget my friends who passed away prematurely, Dr. Ibrahim Ansara from Grenoble (Fr) and Dr. Gunter Effenberg from MSIT Stuttgart (Ge). They were very well versed in phase diagram studies, and we enjoyed all discussions we had, especially in the «Tagungsstätte Schloss Ringberg (Germany)».

Basically, this book is based on the courses I've made during ten years at the University of Montpellier at the levels of Master and Doctorate, during which I was able to observe and analyze the student's reactions to this part of science.

Thermodynamics of materials is a particular science. It always joins strongly physics to chemistry. The proven bestsellers, Alan Prince (Alloy Phase Equilibria) and Mats Hillert (Phase Equilibria, Phase Diagrams and Phase Transformations: Their Thermodynamic Basis), are consistently praised as the most clearly written books available for improving the knowledge of materials.

Some other references to excellent books will be cited as well in the text.

General Introduction

Basis and New Trends in Thermodynamics and Phase Diagrams

The purpose of this book concerns the presentation of modern aspects of the study of phase diagrams and equilibria. Currently, the description of phase equilibria in the twenty-first century requires a multidisciplinary approach linking crystallography and thermodynamics and integrating quantum mechanics.

The history of one of the most exciting sciences of material classes can be traced back to the nineteenth century when J. Willard Gibbs (October 1875–May 1876) (1) discovered the basis of «Equilibrium of Heterogeneous Substances» and published it in the journal: Transactions of the Connecticut Academy of Arts and Sciences. This remarkable first paper developed the theory of chemical thermodynamics and provided the basic theory for the development of this part of physical chemistry.

In the field of metallurgy, the Gibbs theory application provided the basement of the knowledge in multicomponent alloy phase equilibria, and this feature entails the evolution of phase diagrams and applications during the twentieth century. In the twenty-first century, with the progress of experimental determinations and the rise of advanced calculation methodology, the application of the CALPHAD method is presently generalized in alloys and ceramics while in polymer systems the work is not well established. The properties of phase relationships were predicted by simple analysis of the thermal properties of mixtures with good accuracy. More recently, those properties were investigated with modern tools such as physical measurements (electron microscopies, Castaing microprobe and X-Ray diffraction) and quantum mechanics.

The plotting of phase diagrams of two- and three-component systems is not so easy. In both coordinate systems (composition/temperature and/or other coordinates, which are thermodynamic parameters of the phases), the total pressure in the system, molar volume, chemical potentials of components, etc. This is the more pertinent approach to consider and classify.

This book does not claim a replacement for already published books (of quality!). Its objective is to make an analysis of this part of science at a level such that students or even researchers have some difficulties in approaching certain phenomena and experiences in phase equilibria. It is divided into five chapters. Two of them state what is necessary to know in crystallography and thermodynamics in order to apply them to the matter of phase diagrams. Two other chapters are more constructed as guides to the application of the CALPHAD method. They are a gateway to more technical works. Finally, I will present, in a fifth chapter, a small contribution to quantum calculations. But in such a case and to go further, it will be really necessary to use more specialized books.

Contents

Chapter 1
General Considerations on Crystallography: Elements of Crystallography and Their Use in Thermodynamics of Phases

1.1 Introduction

Crystallography is the science that studies the atomic structures of materials. First of all, it is an application of vector geometry in solid-state physics. Moreover in the domain of solid-state chemistry, the word «Crystallochemistry» is often used. It concerns the field of crystallography applications in chemistry which describe the crystalline solid phases by taking into account the nature of atoms, ions, molecules, coordination polyhedra and bonding inside the crystal structure.

The relationships between structure and property and the classification of crystal structures result from this part of physics. The experimental approach of crystallography uses a large panel of experimental determinations such as diffraction and diffusion of X-rays, neutrons and electrons by crystals, but this is not the subject of this book. The reader can view these problems in some related books like [1–5].

This chapter is devoted to the geometric description of the various crystal structures, in particular those concerning the compact staking and the coordination polyhedra in the structures which are the main application in the matter of this book. In the field of phase diagrams application. The concepts of isomorphism and polymorphism will be presented.

From a crystallographic point of view, crystals are described as isomorphous when they are closely similar in shape. The law of isomorphism was discovered by Eilhard Mitscherlich in 1819, working in the laboratory of Heinrich Friedrich Link in Berlin, and he notes that some compounds studied in this laboratory crystallize in the same form [6]. Working on carbonate and calcite crystals, he notes that the angles between the faces of calcite crystals vary with temperature and do not change in the same way in all directions. Investigating the two crystalline forms of sulfur led him to note in 1826, by analogy, that calcite and aragonite have the same chemical formula, but they don't have the same crystalline form, and therefore they show a polymorphism: this is the name he gives to this property of certain crystals.

© The Author(s), under exclusive license to Springer Nature Switzerland AG 2024
J.-C. Tedenac, *Thermodynamics of Crystalline Materials*,
SpringerBriefs in Materials, https://doi.org/10.1007/978-3-030-99027-5_1

1.2 Isomorphism and Phase Transitions

These transformations are classified into four categories [7]. In the case of one compound and depending on temperature and pressure, one can observe one or more phase transformations.

Isomorphism of the first kind: the two phases have a very close chemical formula, the same number of atoms and the same valence between corresponding atoms. One example is the olivine type: Mg_2SiO_4–Fe_2SiO_4.

Second species isomorphism: the ions have a near ionic radius but different typical valence (i.e. one unit).

Isomorphism of the third kind: the change in the charge of one atomic position following the replacement of one ion by another is compensated by a second replacement mobilizing ions of opposite charge (coupled cation–anion replacement). This type of isomorphism is very rare among natural compounds. Example: sellaite MgF_2–rutile TiO_2.

Interstitial isomorphism: it concerns the association of ion-vacancy (wide), which is replacing an ion and the associated wide due to valence balance. Example: in the tridymite structure, SiO_2 can be written as $VASiSiO_4$, or $XSiSiO_4$ (X = additional element in place of vacancy); the trinepheline $NaAlSiO_4$(VA = wide or vacancy) is shown in Fig. 1.1a, b.

In materials science, polymorphism is the ability of a solid material which can exist in more than one form and in crystal structure. Polymorphism can potentially be found in any crystalline material including polymers, minerals, hybrid materials and metals, and is related to allotropy, which refers to the structure of chemical elements. In a crystal, the shape was defined in the past by the measurement of the angles between crystal faces with a goniometer. Presently, it is known that isomorphous crystals

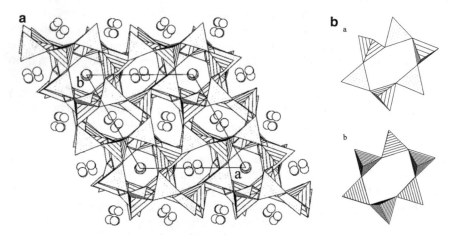

Fig. 1.1 a Crystal structure of trinepheline viewed down *c axis*. Open circles are Na cations. **b** Single six-membered rings of (hight) trinepheline and (below) *F*1-tridymite

should belong to the same space group, and we will discuss the structural phase transitions in the crystals, which is very important in the scope of phase modeling.

The geometric crystallography which is presented deals with the symmetry in crystals. The theory of point symmetry groups is treated, but it can also be used in a large field for isolated molecules in a compound. In the case of atomic packing, the crystal morphology and the space groups are presented.

1.2.1 Reconstructive Transitions

As atoms and ions are changing sites in the structure, reconstructive transitions need the energy to occur: they are first-order phase transitions. If the change of the physical conditions (temperature, pressure) is abrupt, the initial phase can keep its structure out of its field of stability: it is a "metastable region". Then, the structure tends to be transformed into that which corresponds to the physical conditions, but the time of transformation can be important, possibly infinite. Example: the transformation of diamond into graphite.

Sometimes, a structure does not have any stability field in special T, P: it is called a "metastable phase"; it occurs only as an intermediate phase between two other phases, one metastable and the other stable.

These transformations are as follows.

The displacement transitions are second-order phase transitions but sometimes not only, depending on the symmetry changes. As they occur as a result of a simple deformation of the structure, it is not possible to freeze the initial structure out of its stability field. This problem can be evaluated by ab initio calculations, as now some efficient tools are available.

Example: the α-quartz transformation occurs as a result of a simple rotation of the [SiO4]4-tetrahedra which share their vertices, without any chemical bond breaking. Therefore, there is no metastable region for β quartz, which thermodynamically does not exist under ambient conditions, Fig. 1.2.

1.2.2 Polytypism

A particular case of polymorphism is that of polytypism, where different structures are formed by stacking a single module (in most cases, multiple layers or sheets) of structure and composition (almost) identical. The polytypes differ on the mode of stacking (translation and/or rotation) of the module along one direction: the two lattice parameters in the plane of the module are common to all polytypes, while the third differs; see Fig. 1.3.

Fig. 1.2 The alpha-quartz structure. The change to beta phase is obtained by rotation and mirror

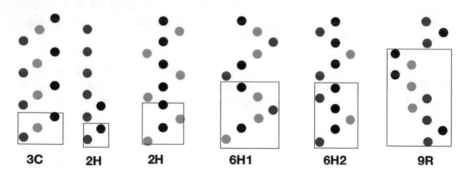

Fig. 1.3 The different polytypes in SiC according to the Ref. [9]

Polytypes are distinguished by specific symbols. The most used symbols are those of Ramsdell, which associate the number of layers in the period of the polytype with the symbol corresponding to the crystalline or reticular system. Figure 1.3 shows the SiC polytypes according to Ramsdell notation which was denied in [8].

1.3   The Periodic Network of Atoms in a Structure: Crystal Structure

A crystalline solid is constituted by the periodic repetition in the three dimensions of space of an atomic or molecular motif, called lattice. The periodicity of the structure of a crystal is therefore represented by a set of regular point arrangements. This set is

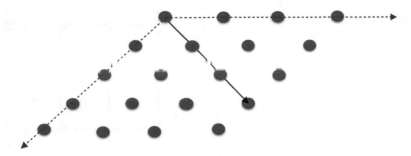

Fig. 1.4 Atom stacking in a bidimentional description

called a crystal lattice and the constituting points are called lattice nodes. A primitive structure is shown in Fig. 1.4.

An overview of the crystallographic properties of perfect crystals can be seen in the references [9]. The crystal which is defined in crystallography is called a "perfect crystal" but in normal life, this stacking is affected by some defects and impurities and leads to "normal" crystal. This problem is described in reference [10]. The crystallographic model is used to represent the structure of all existing crystalline materials. This model considers that a crystal is an ordered and infinite stacking of atoms, ions or molecules; such stacking in two dimensions is shown in Fig. 1.5.

The ideal crystal is a solid with a structure consisting of atoms ordered in a periodic and symmetric lattice. A lattice is an infinite array of points in space, in which each point has identical surroundings to all others. It has symmetry properties with direct and inverse rotation axes, mirrors, planes and center of symmetry.

The elementary lattice is described as the smallest crystalline volume built on the three shortest independent translations of the crystal. It is defined by three vectors generating six parameters: the three lengths of the vectors (a, b, c) and the three

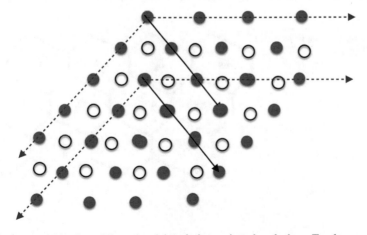

Fig. 1.5 Atom stacking in a tridimentional description projected on the base. Two layers are shown

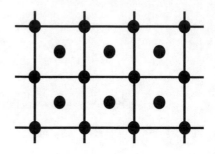

Fig. 1.6 A basic tridimentional cell projection on the xOy plane. Primitive cell

angles between these vectors α, β, γ. They are not colinear. In Fig. 1.6, a lattice in projection on the basic plane is shown. A primitive translation is represented by the vector:

$$t_n = n_1 * a + n_2 * b + n_3 * c \tag{1.1}$$

In this figure, the primitive translation vector is $t_n = 3a + 3b + 2c$. Obviously, the n_i are integers and the vectors a, b and c are not colinear.

A primitive cell is defined as the smallest possible cell. This definition is also connected to the notion of the Wigner–Seitz cell which is useful to understand the definition of the reciprocal lattice. It is the smallest possible primitive cell, and it consists of one lattice point and all the surrounding space closer to it, Fig. 1.7.

The construction of the Wigner–Seitz cell in the reciprocal lattice delivers the first Brillouin zone, Fig. 1.7.

The Wigner–Seitz cell is an elemental cell (i.e. the smallest volume of the crystal lattice that contains all the information). It was named after physicists Eugene Wigner and Frederick Seitz.

It is constructed as the region of space that is closest to one network node than any other node. It is therefore a Voronoi diagram [11].

Fig. 1.7 Construction of a Wigner–Seitz cell

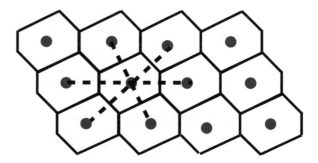

1.3.1   Nodes

A node in the structure is an identified point where an atom or vacancy can be situated. It is characterized by a particular (or local) symmetry named point symmetry.1.3.2 Networks and vectors.

An elementary (or primitive) lattice is a minimal volume that contains only one node in the network. The periodic repetition of this lattice (the specie on the node!) in the three dimensions of space is sufficient to reproduce the entire network and structure. Often, for reasons of convenience or to make the symmetry appear better, the crystal is used to describe a multiple lattice, containing several nodes, which is therefore not elementary: it is the conventional lattice. The multiplicity of the lattice is then defined by the number of nodes it contains.

1.3.2   Lattice [12–13]

A lattice is defined by the three basic vectors. The primitive translation vector is

$$tn = n1\ a + n2\ b + n3\ c$$

Figure 1.8 The vectors a, b and c have the same origin, and they are not coplanar. The lattice parameters are generally used as the lengths of the vectors and angles in between: a, b and c and their angles α, β and γ. The choice of these three vectors is not unique, it is therefore possible to define several elementary cells which can more or less well show the symmetry of the lattice, but usually the crystalline lattice is defined as more symmetric as it is shown in Fig. 1.9.

Fig. 1.8 Representation of a primitive cell with nodes and axes

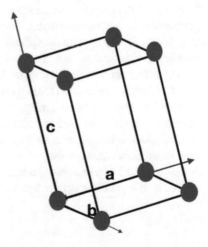

Fig. 1.9 The primitive cubic cell

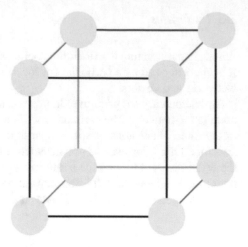

1.3.3   The Bravais Lattice

The Bravais lattices were defined in 1848 by the French scientist Auguste Bravais. The Bravais arrays are classified into 14 lattice types in three dimensions (5 types in two dimensions) and represent the periodicity of the crystal structure. This is obtained from a minimal set of atoms occupying the asymmetric unit, repeated in space according to the operations of the space group of the crystal. All the crystalline materials have a periodicity corresponding to one of these networks (the case of quasi-crystals is a different thing, and it is not treated in this chapter). The 14 Bravais arrays are the expansion of the 7 primitive cells of crystals. A Bravais lattice is a network of nodes obtained by translations following basic vectors from a single node. All primitive lattices are shown in Fig. 1.10. The initial number of Bravais lattices is 7, corresponding to the combination of seven primitive cells; five is the number of primitive cells and two are trigonal or hexagonal (when taking into account that fact and due to symmetry considerations, both these cells are equivalent, and the total number obviously falls to 6. These cells are defined by the label P (Primitive cell.).

These lattices turn into 14 types when considering that other lattice points are added at certain places in as lattice: centered or face-centered. After centering indeed new lattices appear as new arrangements. It results in eight centered lattices (body-centered, one face-centered and all faces-centered), and new symbols are added: I, F, A, B, C and only those (because there is no other way of disposing points in the space). One more thing concerns the hexagonal lattice, after centering the hexagonal lattice can be regarded as a primitive rhombohedral one. From the various combinations of elements of crystalline symmetry, there are 32 symmetry classes.

Bravais lattice	Parameters	Simple (P)	Volume centered (I)	Base centered (C)	Face centered (F)
Triclinic	$a_1 \neq a_2 \neq a_3$ $\alpha_{12} \neq \alpha_{23} \neq \alpha_{31}$				
Monoclinic	$a_1 \neq a_2 \neq a_3$ $\alpha_{23} = \alpha_{31} = 90°$ $\alpha_{12} \neq 90°$				
Orthorhombic	$a_1 \neq a_2 \neq a_3$ $\alpha_{12} = \alpha_{23} = \alpha_{31} = 90°$				
Tetragonal	$a_1 = a_2 \neq a_3$ $\alpha_{12} = \alpha_{23} = \alpha_{31} = 90°$				
Trigonal	$a_1 = a_2 = a_3$ $\alpha_{12} = \alpha_{23} = \alpha_{31} < 120°$				
Cubic	$a_1 = a_2 = a_3$ $\alpha_{12} = \alpha_{23} = \alpha_{31} = 90°$				
Hexagonal	$a_1 = a_2 \neq a_3$ $\alpha_{12} = 120°$ $\alpha_{23} = \alpha_{31} = 90°$				

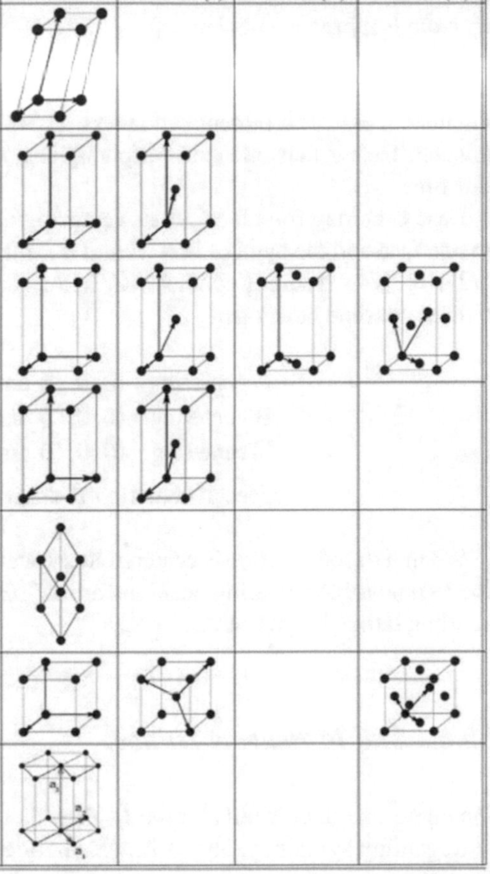

Fig. 1.10 The 14 Bravais lattices

1.3.4 Lattice Centering and Face Centering (See Fig. 1.9)

Six primitive lattices are defined by the assignment of the reference axes in relation to the rotational symmetry. In such a case, the addition of new points can change the symmetry and undergo new Bravais lattices. The required condition is that adding a new point does not change the symmetry. Then this point must be in the center of the cell. This change entails a different cell: centered cubic, (C). Another way is to put the point at the center of faces and in such a case the cell is all faces centered (F) or one face centered (A, B, C for each possible face in a cell). Then it results as follows.

Body centering. This lattice is obtained by placing an additional point at the end of the vector, $t = a/2*x + b/2*y + c/2*z$, meaning that an additional point is at the center of the lattice. It is a unit cell named I containing two points (0, 0, 0) and (1/2, ½, ½) where an atom can be placed. In this case, the multiplicity is 2, indicating that the atom is reproduced twice.

Face centering. By placing an additional point at the center of each face, at the end of the vectors, t = {a/2 + b/2}*x + {a/2 + c/2}*y + {b/2 + c/2}*z, a new lattice is obtained. It is named F (centered faces). This cell contains four lattice points in one unit cell. Then in that case the multiplicity is 4, indicating that the atom is reproduced four times.

Base centering (one face). Base centering is another way for centering obtained on one face and the symbol is A, B and C. If the centering is on a basic plane as ab (a/2 + b/2), the resulting symbol is C. In the same way, one obtains A and B leading to three possible positions:

$$A \text{ centering} : (0, 0, 0) \text{ and } (0, 1/2, 1/2)$$
$$B \text{ centering} \quad (0, 0, 0) \text{ and } (1/2, 0, 1/2)$$
$$C \text{ centering} \quad (0, 0, 0) \text{ and } (1/2, 1/2, 0)$$

Special centering R as rhombohedral.

When a trigonal lattice is centered in a special way, it gives a rhombohedral cell. The two possible centering positions are ±(2/3, 1/3, 1/3) and ±(1/3, 2/3, 1/3). The resulting lattice is symbolized by R.

1.3.5 The 14 Bravais Lattices

The conventional unit cells for the 14 Bravais lattices are shown in Fig. 1.9.

Regarding symmetry, all the Bravais lattices can be changed to different primitive cells but have the same symmetry. This is why understanding symmetry is so important.

1.4 Symmetry in Crystals

1.4.1 Group of Symmetry and Lattice Symmetry

The symmetry in crystallography is divided into two classes: group symmetry and point symmetry.

Definition of a point group. A point group contains the operations of symmetry lattices which leave the asymmetric unit unchanged. Then the symmetry of the atomic structure of a crystal is described by a space group which leaves the symmetry invariant.

Definition of a group symmetry. They are classified into holohedry and meridry groups, depending on the description of the complete symmetry of the network, and sometimes they are subgroups of them. The existence of a periodic network is resulted to the order of the rotation symmetry which in two and three dimensions are limited to characters 1, 2, 3, 4 and 6, whereas these restrictions are not applied to isolated molecules.

This issue is a more general mathematical problem. It corresponds to the analysis of the orthogonal representation of a network (lattice). A lattice is the equivalent of a vector space, with the difference that scalars are integers. The orthogonal group is the linear representation group retaining distances and angles. The space group of a crystal consists of the set of symmetries which concerns the crystal structure. The set of isometries leaves the structure invariant. Therefore, it is a group in the mathematical sense of the term.

The point groups.

There are 32 point groups. These point operations in the crystal are of two types.

Operations of the first kind do not change the chirality of the object on which they are acting; they are pure rotations (Fig. 1.10).

The first element of symmetry is identity. It is an operation denoted by 1 which associates with each point P of space the point P itself as an image. It is an operation without a geometric effect. Reflection (Fig. 1.9) ikeeps the distances; it is an isometry. It does not preserve the chirality of the objects. Chirality is also defined by reflection on a plane: a chiral object is an object that is not superimposed on its image in a mirror. The second symmetry element changing geometry is the rotation of angle θ. It is an operation which associates at any point P of space an image point P'which is rotated by the angle θ with respect to the axis of the rotation. The angle of rotation is expressed in degrees. The rotation is in the direction of the axis and in the plane containing the point P which is perpendicular to the axis. The fixed points in a rotation constitute the axis of the rotation. Figure 1.10 shows the application of a rotation of angle γ around the z-direction on the x- and y-axes in a Cartesian coordinate system. Rotation is an isometry: it keeps the distances. In particular, the distance from point P to the axis of rotation is the same as the distance from its image P' to the axis. On the other hand, the rotation preserves the chirality of objects, as shown in Fig. 1.10 b and c: the rotating image of a levorotatory object is a levorotatory object. The order n of a rotation is defined as the number of successive rotations to be applied at point P returning to the initial position. For example, a rotation of 180° around an axis is a rotation of order 2: the application of P of two successive rotations of 180° around the same axis restores P. If the rotation angle θ is a divider of 360°, the order of the rotation is n = 360/θ. The rotation of order 1 is a rotation of angle 360° and is therefore the identity. A rotation is noted by its order, n. Operations of the second kind change the chirality of the object; they are rotation-inversions, consisting of a rotation followed by an inversion. It is a symmetry, with respect to a point, called an inversion center (see Fig. 1.10c).

For each operation of the first kind, an operation of the second kind can be associated with transforming the barycenter of the object as the operation of the first kind

Fig. 1.11 Réflexion axis.
The pattern A is obtained
from top to bottom by a
reflexion symmetry. It'
include a symmetry centre

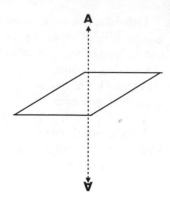

associated with it. When the object is not a chiral one, the result of the application of these two operations of symmetry is identical.

In two-dimensional space, the reflection is made with respect to a line called a "mirror line". Rotation is an isometry: it keeps the distances. In particular, the distance from point P to the axis of rotation is the same as the distance from its image (P) to the axis.

Inversion.

In crystallography, inversion is noted by one bar over the number() and corresponds to the central symmetry in mathematics. It is performed with respect to an inversion center. The fixed point of inversion is this inversion center.

Inversion associates at every point P of space an image point P' which is symmetrical to it with respect to the inversion center situated on the origin O such that O is at the center of the segment [PP']. If the point P has coordinates (x, y, z), its image P' then has coordinates ($-x$, $-y$, and $-z$), (Fig. 1.11).

Roto-inversion.

This is composed of a rotation of order n around an axis and inversion with respect to a point of the axis. These two symmetry operations switch: the order in which they are performed does not change the result of the roto-inversion. There is only one fixed point of roto-inversion, which is the inversion center used for the compound. Roto-inversion is noted as shown in Table 1.

The principle of coordination volume. When considering a central atom associated with its surrounders, it can be defined as a coordination polyhedron around it. This polyhedron is of great interest in the thermodynamic modeling of the phases by using a sub-lattice model. Then it is necessary to take into account the symmetry of this model which is a point symmetry.

1.4.2   Space Group: Plans, Directions and Indices

The space groups.

Each space group results from the combination of a Bravais lattice and a point symmetry group: All symmetry of the structure results from the product of a translation of the lattice and a transformation of the point group. In terms of representation, the Hermann–Mauguin notation is used to represent a space group, and Schoenflies notations are also sometimes used by a few scientists [12]. Finally, there are only 250 space groups where all crystals can be represented and only those existing for all crystals. They are summarized in the International Tables of Crystallography published by the Union of Crystallography [13]; in volume A, each space group and its operations of symmetry are represented graphically and mathematically.

The set of space groups results from the combination of a basic unit with its point operations of symmetry (which are reflection, rotation and inversion), to which are added operations of translation, translation in the plane or combined with reflection or rotation (glide planes and helicoïdal axis).

However, the number of distinct groups is less than that of the combinations, some being isomorphic, which leads to the same space group. This result can be demonstrated mathematically by the group theory.

The translation operations are done according to the basic vectors of the lattice, which passes from one polyhedron to the neighboring polyhedron. The translations can be combined with reflections and rotations and entail new symmetry operations as follows.

Helicoidal axis: it's composed of a rotation along an axis, combined with a translation in the same direction, Fig. 1.12. The translation amplitude is based on a fraction of the basic vectors. They are denoted by a number n describing the degree of rotation, where n is the number of times the rotation must be applied to obtain the identity (6 thus represents, for example, a rotation of one-sixth in turn, i.e. $2\pi/6$). The degree of translation is then denoted by an index which indicates to which fraction of the vector of the lattice corresponds the translation. In general, the helicoidal axis represents a rotation of $2\pi/n$ followed by a translation of p/n of the vector of the array parallel to the axis.

Fig. 1.12 Rotation axis. From a symmetry center the image is obtained by rotation around the z axis

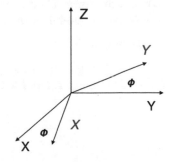

Glides: a sliding reflection, that is to say, a reflection followed by a translation parallel to the plane, as defined in the following table.

The set of space groups results from the combination of a basic unit (or pattern) with point operations of symmetry (reflection, rotation and inversion), to which are added operations of translation, translation in the plane or combined with reflection or rotation.

However, the number of distinct groups is less than that of the combinations, some being isomorphic, that is, leading to the same space group. This result can be demonstrated mathematically by the group theory, [12].

In a space group, different elements of symmetry of the same dimensionality can coexist in parallel orientation. For example, axis 21 must be parallel to axes 2; and mirrors of type m must be parallel to mirrors of type a. In the space group symbol, the choice of the representative element follows a priority order, which is indicated as follows:

The non-slip axes have priority over the helicoïdal axes.

The priority in the choice of the representative mirror is $m > e > a > b > c > n > d$.

However, there are a few exceptions. For example, the groups I222 and I212121 contain axes 21 parallel to axes 2, but in the first group the three 2 axes have a common intersection as well as the three axes 21, while in the second group this is not the case. The priority rule does not apply here, otherwise the two groups would have the same symbol.

1.4.3   The Nomenclature of the Crystal Structures

The set of space groups results from the combination of the base unit with point operations of symmetry (reflection, rotation and inversion), to which are added operations of translation, translation in the plane or combined with reflection or rotation.

However, the number of distinct groups is less than that of the combinations, some being isomorphous, that is to say, leading to the same group of space. This result can be demonstrated mathematically by group theory.

The first nomenclature of crystals is the space groups [12], but there is any indication in such nomenclature about the crystal chemistry. In this case, the nomenclature is described in several manners.

– The prototype.

Prototype is the name of the phase for which this crystal structure was determined for the first time (e.g. NaCl or CsCl).

The Strukturbericht nomenclature. This nomenclature is originally coming from a German notation and it gives letters and numbers within a sequence of classification. This nomenclature doesn't cover the recent determinations because it was stopped around the year 1950. Then a new nomenclature was created and used for the further determined structure.

The "Pearson" symbol is a description of the structure organized by a few comprehensive symbols. Two letters are used, one minuscule indicating the Bravais lattice, (c, t, h, o, m, ..) and one majuscule showing the nature of the lattice (P, C, F, I, R) with an integer representing the atom number in the unit cell.

In some cases, the mineral name is used such as perovskite or spinel.

Sometimes, a little change from this nomenclature is used (for example, ordered BCC for CsCl or ordered FCC for tP2). In our opinion, this representation is not really correct by introducing confusion in the description of the Bravais lattices, but it is very often used.

1.5 Crystallochemistry

Crystallochemistry (from the Greek κρύσταλλος and χημεία) is the study of the relationship between the chemical composition of crystalline materials and their crystal structures, as well as their effects on physical properties. It thus makes the link between crystallography, chemistry of the solid and physics of condensed matter. A neighboring branch of crystal chemistry is the chemistry of complexes which out of this book matters.

This concept was defined by Victor Goldschmidt and Fritz Laves (AA), postulating about the space filling for a stable crystal structure of minimal reticular energy (with the restriction as atoms or ions are considered geometrically as hard balls). It consists of a few simple principles [14].

- The space filling principle: the atoms or ions are stacked as compactly as possible.
- The principle of symmetry: the crystal possesses the highest possible symmetry according to the symmetry group.
- The principle of interaction: each atom or ion is surrounded by as many neighbors as possible.

In addition, the atomic or ionic radii (which are depending on the type of bond) have an influence on the structure. Generally, atom stacking is made by the bigger atoms (or ions) and the others fill the other sites. In the case of perovskites, for example, the crystal structure can be described as an FCC stacking of oxygen atoms, the other smaller atoms occupying the octahedral sites and tetrahedral sites.

- The nature of the chemical bonds in a crystal may be described with a predominant type of bond or by groups of atoms or stable and isolated complexes, included in a larger structure. The pyrite compound FeS_2 or the skutterudite $CoSb_3$ is an

example of a mixture of bonding (covalent bond between sulfur atoms or antimony in the second example, ionic or metallic bond for those including the p elements).
- The Pauling rules are applied to ionic crystals. In these structures the ionic bonds predominate.

1.6   Defects in Lattices and Structures

From the crystal description, the structures presented above can be considered as ideal structures which are not really observed by experience. In all real cases, these structures contain many defects which are caused by the different experimental procedures used in the compound synthesis. Generally, crystalline defects are the origin of some interesting properties of the material, such as electrical conductivity and especially electronic, plastic deformation, breaking strength and color centers.

Four types of defects exist: The point defects, which are of 0 dimension. They are named vacancy, interstitial atoms (self- or hetero-interstitial) and anti-sites.

They concern a single node, and depending on the characteristics of the foreign atom they disturb only a few neighboring nodes. A vacancy is defined by the absence of a few atoms in their nodes and they are statistically distributed. The interstitial atoms are the atoms which are placed in the possible wide nodes of the structure; they can be self-interstitial (same atom) or hetero interstitial (sometimes H, B, C, O). Concerning the anti-site defects, one should note that usually they exist when the difference between the atoms in the structure is not so large that exchange between atoms on different sites is possible.

1.7   Linear and Volumic Defects

The linear defects of 1 dimension: dislocations (screw or corner); they concern only the nodes situated on a line and disturb only a few neighboring nodes. Surface defects of 2 dimensions: grain boundaries, phase boundaries, twinning, stacking faults and antiphase boundaries. Concerning the main application of this chapter, we will describe only the point defects and those concerning their involvement in the thermodynamic stability of phases. It means that they are realistic when they are acting on physical properties and structure, for example in a metallic or intermetallic alloy. Considering the charged defects (anions or cations or charged vacancies) is not realistic in the metallic and intermetallic compounds; they are only important in the ionic compounds. Point defects can be divided into three major categories depending on their geometry: vacancies or point deficiencies, interstitial atoms, substitution atoms and complex point defects. Point defects are chemical species which can give rise to chemical reactions, so they are defined by a chemical potential which affects the value of the energy of the formation of compounds. Also, they can associate, for example, opposed charge defect attraction and bonding in iono-covalent solids. The notations of Kröger and Vinik [9] are used in such a case to designate them. The point defects confer specific properties on crystals, such as color (color centers), motion of species in the crystal due to the diffusivity, electric conduction in particular in semiconductors and some crystalline oxides. When the defect concentration is low, they can be considered as isolated. Therefore the creation of defect requires energy which only depends on the environment, and which is the same everywhere in the crystal. The defects can be associated and the mechanisms involved are intrinsic defects (Schottky or Frenkel defects), electron–hole pairs and redox conditions for oxides. An intrinsic defect such as Schottky or Frenkel is the association of an anionic and cationic vacancy. It concerns only the ionic compounds, and one can consider that this depends only on the electronegativity difference of the elements involved. Also, clusters of defects are obtained when more than two defects are associated (one often uses the term cluster). Clusters generally appear for defect concentrations greater than 10–2 defects/mesh (one defect every 100 mesh).

1.8 Lying Crystallography and Thermodynamics

The knowledge of the crystal structure of a compound is very important for modeling the phase by using statistical thermodynamics, in particular in the so-called "Compound Energy formalism". By considering that all atoms are in equivalent positions and as they should have they have the same surrounding atoms.

A set of equivalent positions are treated as one sub-lattice, then the structure must be presented as the sum of different sub-lattices with the same Wyckoff positions. In the scope of the "Compound Energy formalism", it is very important to understand the crystallochemistry of compounds as it was described before, particularly the description of the phases with the use of coordination polyhedra. This treatment of the modeling of the phases by using the sub-lattices theory was described in different recent books on thermodynamics. Finally, and this is very important, by the consideration of Wyckoff positions, the defects in structure and the surrounding of elements, one can decrease the number of sub-lattices in the treatment of phases and go to a smarter description. This problem will be discussed in the next chapters. (Figs. 1.13 and 1.14).

Fig. 1.13 Rotation (order 2) and reflexion

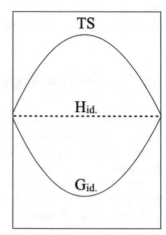

Fig. 1.14 Variation of the thermodynamic functions of mixing for a regular solution comparatively to a ideal mixing

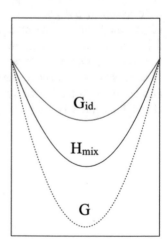

References

1. D.E. Sands, *Introduction to Crystallography* (Dover publications Mineola, NY, 1994)
2. M.D. Graef, M.E. McHenry, *Structure of Materials: An Introduction to Crystallography, Diffraction and Symmetry*, 2nd edn. (Cambridge University press, NY, 2012)
3. C. Giacovazzo, H.L. Monaco, B. Viterbo, F. Scodardi, G. Gilli, G.Zanotti, M. Catti, *Fundamentals of Crystallography* (Oxford University Press, 2011)
4. C. Hammond, The basics of crystallography and diffraction, in international Union of crystallography, 3rd edition (Oxford University Press for the International Union of Crystallography, Oxford, UK, 2009)6- D. Schwarzenbach, «Cristallographie», (in french », Presses Polytechniques et Universitaires Romandes, 1st edition , 1993.
5. D. Schwarzenbach, Cristallographie, (in french), 1st edn (Presses Polytechniques et Universitaires Romandes, 1993)
6. E. Mitscherlich, Premier Mémoire sur l'Identité de la forme cristalline chez plusieurs substances différentes, et sur le rapport de cette forme avec le nombre des atomes élémentaires dans les cristaux, (in french). Annales de Chimie et de Physique t. XIV **1820**, 172–190
7. A.F. Wells, *Structural Inorganic Chemistry*, 5th ed. (Oxford University Press, 1984)
8. A. Guinier et al., Nomenclature of polytype structures, nomenclature of polytype structures, report of the international union of crystallography ad-hoc committee on the nomenclature of disordered, modulated and polytype structures. Acta Cryst. A40, 399–340 (1984)
9. C. Kittel, *Introduction to Solid State Physics*, 8th edn (Wiley, USA)
10. K. Kosuge, *Chemistry of Non-stoichiometric Compounds.* (Oxford University Press, 1993)
11. P.A. Burrough, R.A. McDonnell, C.D. Lloyd. 8.11 Nearest neighbours: Thiessen (Dirichlet/Voroni) polygons. Principles of Geographical Information Systems. (Oxford University Press, 2015)
12. A. Sands, E. Donald, *Crystal Systems and Geometry. Introduction to Crystallography* (Dover Publications, Inc., Mineola, New York, 1993), p. 165
13. M.I. Aroyo (ed.), *International Tables for Crystallography, Volume A: Space-group Symmetry* (2016)
14. V. Kahlenberg, H. Bohm, Crystal structure of hexagonal trinepheline—a new synthetic NaAlSiO4 modification. Am. Mineralog. **83**, 631–637 (1998)

Chapter 2
Thermodynamics of Materials and Basis

2.1 Introduction: The Modern Approach of Phase Diagrams

The main interest in thermodynamics of materials is presently understandable by a global approach to the phase diagrams, phase equilibria and phase stabilities. Phase diagrams are the well-known description of pressure, temperature and phase quantities (p, T, Niφ) of the phase relationships in a system. Phase equilibria describe the phases in a system which are in equilibrium in certain p, T ranges through the Gibbs–Duhem equation (which will be defined below). Phase stabilities can be easily described by the so-called "Free Energy Function" of the phases in a system at 0 K and in the whole range of temperatures. It depends on the lattice stabilities which were defined firstly in the paper of Andersson et al. [1]. Consequently one should understand the slight (but fundamental) differences between properties of the phases. Therefore, taking into account all these characteristics of materials undergo to a modern approach of phase diagrams. The thermodynamics of materials is nowadays mainly described by the CALPHAD method which is globally explained in four excellent books [2–7]. We will show below what is necessary to know in the modern approach to phase equilibria and then the reader can progress in the knowledge of this branch of thermodynamics.

2.2 Basis and History. Heat Quantity, Temperature and System

«Whoever does not know where he comes from cannot know where he is going»—Antonio Gramsci.

Over the ages, the notion of heat has always been associated with various observable chemical processes. This notion is very old since it was associated with cooking, skin tanning, metal melting, etc.

© The Author(s), under exclusive license to Springer Nature Switzerland AG 2024   21
J.-C. Tedenac, *Thermodynamics of Crystalline Materials*,
SpringerBriefs in Materials, https://doi.org/10.1007/978-3-030-99027-5_2

Until the 1600 s, the theory of elements (air, earth, fire) was known. This theory was put to shame in 1661 by Robert Boyle. Then latter, among other scientists of that period, was interested in the phenomenon of combustion, and the theory of the nitroaerian spirit (that is to say, coming from nitre) appeared.

This was the beginning of the explanation of combustion by the intervention of a particular component of the matter: phlogiston. Indeed, in 1697 Georg E. Stahl hypothesized that any combustible substance that contained a constituent named phlogiston escaped from it burning and left an ashes residue. This approach satisfied everyone because, of course, it had been found that, after combustion, the weight of residual ash was less than the weight of matter involved, the difference was due to phlogistics. As a corollary, it was defined that the burning substances were those which contained the greatest quantity of phlogiston, hence this is the relation not yet established with the notion of the quantity of heat.

In 1756, the Russian scientist Lomonosov dismissed this theory by arguing about an increase in the weight of metals during their combustion. Indeed, this was contrary to the theory of phlogiston, so he didn't still offer any relevant explanation but the right way was actually defined.

It was not until the discovery of oxygen (Carl Wilhelm Scheele) to find the solution to this thorny affair. Finally, due to the genius of Antoine de Lavoisier, practicing science during his own hours, the theory of phlogiston was definitively questioned and replaced by a theory, more in keeping with the most recent observations. But it must be remembered here that finally there is no frontal opposition between the two notions, because they were different perceptions of the same phenomenon.

Lavoisier establishes the well-known law of mass conservation during a chemical reaction, the formulation of this law being "The total mass of substances in a reaction is equal to the total mass of the reagents involved." This is reflected in some works by the very famous phrase: "Nothing is lost, nothing is created, everything is transformed". This proposition is reprinted by the so-called cycle of Carnot reprinted in Fig. 2.1 with parameters T and P.

It was then gradually realized that these propositions were more in conformity with the experimental observations, and from this period, the theory of phlogiston was gradually abandoned as Lomonosov questioned.

Later, the work of Sadi Carnot in 1814 brought to light and masterly the energetic nature of the heat.

In 1814, Sadi Carnot, in his book "Reflections on the Driving Power of Fire", showed that there is a relationship between the work done by a steam engine and the difference in temperature between the hot and cold sources of the machine. This was the beginning of classical thermodynamics or macroscopic thermodynamics, Fig. 2.2.

All that remained was to demonstrate the equivalence of work and heat. This is what Clausius (1822–1888) did by introducing the thermodynamic functions and the statement of laws or principles. The notion of temperature is born from the notion of hot and cold. For a very long time, heat and temperature have often been confused. It was only by necessity due to his theory that Sadi Carnot made a difference between temperature and energy, heat being a particular form of energy. On the other hand, the notion of thermodynamic temperature is in fact recent. It derives from the zero principle of thermodynamics (or Nernst principle) which will be shown below. It is very important to have a clear understanding of the basic knowledge of fundamental thermodynamics, and this book does not claim to give a thorough knowledge of this discipline but rather to give the codes of access to the specialized works in this field [3, 4].

2.3 Theoretical Background

The development of thermodynamics starts with the definition of Q. It represents the amount of heat flow entering into a closed system, and W is the amount of work done on the system. The concept of work may be actually regarded as the change in the surroundings and which changes are made in the system.

The first law of thermodynamics is related to the law of conservation of energy. It says that energy cannot be created or destroyed. As a consequence, if a system receives at the same time an amount of heat (Q), and if a work (W) is done, the energy of the system must be changed by quantity $Q + W$. This is quite independent of what happened to the energy inside the system. Finally, the concept of internal energy U has been invented, and therefore the first law of thermodynamics can be formulated as

$$U = Q + W \tag{2.1}$$

In a differential form we will have:

$$dU = dQ + dW, \tag{2.2}$$

It is shown in Fig. 2.2.

It is rather evident that the internal energy of the system is uniquely determined by the state of the system and independent of the route, and it has been established as U is a state variable. It should be emphasized that Q and W are not properties of the system, but they define different ways of interaction with the surroundings. Thus, they could not be state variables. A system can be brought from one state to another by different combinations of exchange in heat and work. So it is possible to bring the system from one state to another by a different route and then let it return to the initial state by a different route; this is due to the fact that these functions are functions of the state. It would thus be possible to get mechanical work out of the system by supplying heat and without any net change of the system.

It is not easy to vary experimentally internal energy U, in a controlled fashion. Thus, we shall often look at U as a state function rather than a state variable. At equilibrium it may, for instance, be convenient to consider U as a function of temperature and pressure because those variables are more easily controlled in the laboratory:

$$U = U(T, P). \tag{2.3}$$

Obviously, the function U is an extensive property and obeys the law of additivity. Consequently, the total value of U of a system is equal to the sum of the different U values of the various parts of the system. Its value does not depend on the path used to go from the initial value to the final value, and due to the heat and work exchanged in the system.

It should be emphasized that the absolute value of U is not defined through the first law, but only changes of U are then defined. One can only consider changes in internal energy because there is no natural zero point. This is why in the past an arbitrary zero point has been chosen.

In the case of compression work on a system under a hydrostatic pressure P, the equations are

$$dW = P(-dV) = -PdV \tag{2.4}$$

$$dU = dQ - PdV. \tag{2.5}$$

So far, this definition of U is used in cases where the system is closed, and the work done on the system is hydrostatic. Consequently, this treatment is always applicable to gases and liquids which are not submitted to shear stresses. In order to be coherent with a fine analysis of the problem, one should emphasize that the treatment of the solid materials requires consideration of non-hydrostatic stresses. By measuring, the content of matter in units of mole equation must be changed as

$$U = -PV + Qm. \tag{2.6}$$

These expressions question the changes according to a shape (or volume) variation.

2.3.1 External State Variables

Thermodynamics is concerned with the state of a single system interacting with the surroundings. The "system" is a part of the world, defined by considering the changes that may occur under varying conditions. The system may be separated by a wall from the surroundings [8]. A surface S is isolated from the surrounding by a wall and contains different domains σi. These domains are contained in the surface and show properties of S, Fig. 2.3. The properties of the wall determine how the system may interact with the surroundings. The wall itself will not usually be regarded as a part of the system but as part of the surroundings, meaning that for the system the wall is immaterial. We shall firstly consider only two kinds of interactions, thermal and mechanical, and they are obviously the more important in the solid state. Then

"Thermodynamics" is the proof that these interactions are of interest. Secondly, it is needed to introduce interactions through the exchange of matter in the form of chemical species. The name "thermochemistry" is sometimes used as an indication of such applications. Let's consider that the name "thermophysical properties" is used for the representation of thermodynamic properties which do not normally involve some change in the content of the chemical species, such as heat capacity, thermal expansivity and compressibility.

One might imagine that the content of matter in the system could vary in a number of ways which are equal to the number of species present in the system. Then the species may react with each other inside the system; this is the matter of chemistry. Therefore, it is convenient to define a set of independent components. By denoting the number of independent components as c and considering at the same time all possible thermal and mechanical interactions with the surroundings, by definition, the state of the system is varying in c + 2 independent variables. For metallic systems, it is usually more convenient to assume that the elements are independent components. For systems containing covalent bonds, it is often convenient to consider a stable molecular specie as a single component. Concerning the systems with a strongly ionic character, the neutral components should be selected as independent components rather than ions and it is necessary to take into account the neutral character of the mixture. Concerning iono-covalent (with localized bonds) systems, it is better to consider these systems as metallics.

What is phase stability? It is a criterion that is spontaneously established from different starting points giving a state of equilibrium I (Fig. 2.4). After a system has reached a state of equilibrium, it is possible, in principle, to measure the values of quantities which are uniquely defined by the state and independent of the history of the system (temperature T, pressure P, volume V and content of each component N_i). Such quantities are called state variables or state functions. Also, it is possible to identify a particular state of equilibrium with the values of a number of state variables. For example, when c + 2 variables are given, the values of all other variables are actually fixed, and equilibrium is really established. There are thus c + 2 independent variables and, after they have been selected and equilibrium has been established, the other are dependent variables. For each application, a certain set is usually the most convenient. For any selection of independent variables, it is possible to change the value of each one, independent of the others, but only if the wall containing the system is open for exchange (exchanges of mechanical work, heat and c components).

The equilibrium state of a system is represented by a point in a c + 2 dimensional diagram. In principle, all points in such a diagram represent possible states of equilibrium although there may be practical difficulties in establishing the states represented by some regions. One can use the diagram to define a state by specifying a point (unique state of equilibrium) or a series of states specifying a line. Such a diagram may be regarded as a state diagram. It does not give any information on the properties of the different phases and systems under consideration. Such information can be added to the diagram. We shall later see that information on the properties can be included in the state diagram. Therefore, in order to show the value of some dependent variable a new axis must be added. For convenience of the representation, it is possible to decrease the number of axes in the c + 2 dimensional state diagram by sectioning at constant values of c + 1 of the independent variables. All the states to be considered will thus be situated along a single axis, which may now be regarded as the state diagram. We may then plot a dependent variable by introducing the second axis. That property is thus represented by a line. We may call such a diagram a property diagram. An example is shown in Fig. 2.1. Of course, we may arbitrarily choose to consider any one of the two axes as the independent variable. The shape of the line is independent of that choice, and it is thus the line itself that represents the property of the system.

In many cases, the content of matter in a system is kept constant and actually to the value 1 of a mole of matter. The wall is only open for the exchange of mechanical work and heat. Such a system is called a "closed system". In other cases, the content of matter may change and, in particular, the composition of the system by which we mean the relative amounts of the various components independent of the size of the system. In materials science, such an open system is called an "alloy system", and its behavior as a function of composition is often shown in so-called phase diagrams.

2.3.2   Internal State Variables

The state variables of the isolated system are of two kinds, and they are called intensive and extensive. Intensives are those that can be defined at each point of the system (temperature T and pressure P). It is important to note that, in particular, T must have the same value at all points in a system at equilibrium. This property gives the special property of intensive variable as potential. There possibly exist some other intensive variables, but as they have different values at different parts of the system, they will not be regarded as potentials.

Volume V, a variable related to mass, is an extensive variable; its value for a system is equal to the sum of its values of all parts of the system. The content of component i, usually denoted by n_i or N_i (and $N_{i\varphi}$ for the phase φ), is actually an extensive variable. Such quantities obey the law of additivity. For a homogeneous system, their values are proportional to the size of the system.

One can imagine variables, which depend upon the size of the system but do not always obey the law of additivity. The use of such variables is complicated and will not be much considered in the matter of this book. The law of additivity will be further discussed.

If the system is contained inside a rigid wall, thermally insulating and impermeable to matter, then all the interactions mentioned are prevented and the system must be regarded as completely closed to interactions with the surroundings. It is often called an isolated system. On the other hand, by changing the properties of the wall we can open the system to exchanges of mechanical work, heat or matter. A system open to all these exchanges may be regarded as a completely open system. We may control the values of c + 2 variables by interactions with the surroundings, and we must regard them as external variables because their values are obviously changed by interaction with the external world through the surroundings.

We do not speak about another variable useful for chemical reactions, the reaction progress rate which was introduced by Th. De Donder (ξ); it doesn't fall under the scope of this book even if it is very important in the field of chemistry and chemical engineering.

2.4 Laws of Thermodynamics

2.4.1 The First Law of Thermodynamics

This law corresponds to the energy conservation principle.

For any thermodynamic system, one can define, with a constant, a function U, called internal energy, having the following properties:

U is a state function (it depends only on the initial and final states of the transformation);

U is extensive;

U is defined for an isolated system.

The change of U during an infinitesimal transformation in a closed system (of fixed composition) satisfies the relation:

$$dEc + dEp + dU = \delta W + \delta Q \tag{2.7}$$

It is very important to consider the impact of a mechanical work of hydrostatic pressure (pi) in the scope of phase diagram studies. Therefore, it is very convenient to define a special state function in the following way:

$$H = U + PV \tag{2.8}$$

It is called enthalpy H.

According to this definition, the first law can then be written as

$$\Delta H = dU + PdV + VdP = dQ + VdP. \tag{2.9}$$

Moreover, the internal energy must depend on the quantity of matter, N, and for one open system subject to compression, we should be able to write

$$dU = dQ - PdV + KdN. \tag{2.10}$$

What is the K parameter?

In order to identify the nature of the parameter K, one shall consider a system that is part of a larger, homogeneous system for which both T and P are uniform. U may be evaluated by starting with an infinitesimal system and extending its boundaries until it encloses the volume V. Since there are no real changes in the system dQ = 0 and P and K are constant, we can integrate from the initial value of U = 0 where the system has no volume, obtaining Hm (the molar enthalpy). Molar quantities will be discussed in Sect. 2.6 (Solutions). The first law can thus be written as

$$dU = dQ + dW + HmdN. \tag{2.11}$$

Since the publication of Gibbs' last paper [9] in the series "On the Equilibrium of Heterogeneous Substances" in 1877, all terms necessary to describe the chemical equilibrium are defined. The chemical potential had been introduced, and the relation describing the different types of phase diagrams (the Gibbs–Duhem Eq. 2.22) had been derived too. Most problems dealt with in equilibrium thermochemistry are those with constant temperature and pressure and where the other work terms, except for the chemical contribution, are usually omitted. It is important to keep this in mind since the entire database derived under these conditions is a Gibbs energy, rather than a Helmholtz, Enthalpy or Internal energy database. Electrochemistry, of course, can only be treated if the electrical work term is also explicitly included. Problems with constant temperature and volume, for example, have thus to be treated in an indirect way, which is, of course, no problem for the computer.

Using the Maxwell relations, one can easily relate the Gibbs energy in its natural variables G(T, P) with the other state functions associated with their own natural variables: the Helmholtz energy, F(T, V), the enthalpy, H(S, P), and the internal energy, U(S, V).

2.4.2 The Second Law of Thermodynamics

The second law of thermodynamics (also known as the second principle of thermodynamics or «Carnot principle») establishes the irreversibility of physical phenomena, especially during heat exchanges. It is a principle of system evolution which was first stated by Sadi Carnot in 1824. It has since been the subject of numerous generalizations and successively the formulations of Clapeyron (1834) and Clausius (1850), Lord Kelvin and Ludwig Boltzmann. In 1873 and Max Planck are known, throughout the nineteenth century and beyond to the present day.

The second principle introduces the function of the entropy state, S, called entropy. S is usually described as the sum of two terms: an exchange term and a creation term. The entropy variation of a system, during any transformation, is described as the equation:

$$\Delta S_{sys.} = \Delta S_{exc.} - \Delta S_{creat.} \tag{2.12}$$

The term creation is always positive or null; it imposes the meaning of the evolution of transformation only for a reversible transformation.

The exchange term in the case of a closed system, delimited by a surface Σ, is expressed as

$$S_{exc.} = \int_{\Sigma} \frac{\delta Q(r, t)}{T(r, t)}$$

(2.13)

$T(r, t)$ is the temperature at the point where the heat is exchanged with the external environment and at the instant t, and $\delta Q(r, t)$ is the heat exchanged with the external environment. Another formulation is possible, as we have seen previously, considering the entropy of the system and the entropy of the external environment. This formulation is fully compatible with the previous one:

$$\Delta S_{total} = \Delta S_{creation} = \Delta S_{syst.} - \Delta S_{ext.}$$

(2.14)

If we stay on the side of the external environment, the sign is reversed according to the rule of signs and therefore, we have the following.

Sexc. corresponds to the entropy exchanged by the system with the external environment. If we place ourselves on the side of the external environment, according to the rule of signs the sign is reversed and therefore

$$\Delta S_{ext.} = -\Delta S_{exc}$$

(2.15)

Then:

$$\Delta S_{syst.} = -\Delta S_{ext.} + \Delta S_{exc.}$$

(2.16)

and

$$S_{creat.} = \Delta S_{sys.} + \Delta S_{ext.}$$

(2.17)

The global entropy variation corresponds to the entropy created and is equal to the sum of the entropy variations of the system and the external environment. It is always positive in the case of real irreversible transformations. On the other hand, in the ideal case of reversible transformations, it is null.

Consider a transformation, at the temperature T, carried out either reversibly or irreversibly. The entropy being a function of state, its variation will be the same for the two paths envisaged. On the other hand, the heat will depend on the path followed because it is not a function of state.

Reversible transformation in which the entropy creation is null:

$$\Delta S^{syst.} = S_{exc.} = \frac{Q}{T}$$

(2.18)

Irreversible transformation in which the entropy creation is positive; it follows:

$$\Delta S^{syst.} = S_{exc.} + S_{creat.} \tag{2.19}$$

then

$$\Delta S^{syst.} = \frac{Q_{irr.}}{T} + S_{creat.} \tag{2.20}$$

it results in

$$\Delta S_{syst.} > \frac{Q_{irr.}}{T} \tag{2.21}$$

This expression was formulated by Clausius. It is still called Clausius' inequality. This is another way of expressing the second principle.

2.4.3   The Third Law of Thermodynamics

The so-called, third principle of thermodynamics is also called the principle of Nernst (1906).

It states that "the Entropy of a perfect crystal at 0 K is equal to zero."

This makes it possible to have a determined value of the entropy (and not "to an additive constant"). This principle is obviously linked to the quantic indiscernability of identical particles.

As it is difficult to attend to this principle, it is much easier to talk about consequences. The heat capacities at constant volume or pressure (Cv and Cp) must tend toward zero when T tends to zero. This is the case for the calorific capacity of crystals (Cv) since $Cv = a \times T3$ (Debye law at low temperature). In the case of metals, when the temperature becomes very low, the contribution of free electrons must be taken into account and the electron capacitance is $Cv, elec = \gamma \times (T/TF)$ where TF is the Fermi temperature and the electronic contribution also tends to zero when $T \rightarrow 0$ K.

One cannot reach absolute zero. If we consider that the right variable to consider the temperature is—(1/T) or —1/(kT), then it means that this variable tends toward infinity, which of course is never attainable. In fact, the variable—(1/T) is the intensive parameter associated with energy U.

Nevertheless, since the temperature is the intensive variable associated with entropy S, in the field of statistical thermodynamics, in special cases, negative temperatures can be obtained (but in this case it has nothing to do with the thermal notion of heat and the temperature T must be considered as the intensive parameter associated with S).

2.5 Entropy

In classical thermodynamics, entropy is an extensive state function (as it was explained here, it is dependent on the mass or volume of matter); this function was introduced in 1865 by Rudolf Clausius within the description of the second principle of thermodynamics (after the work of Sadi Carnot). Clausius showed that the ratio Q/T is less than or equal to the variation of a state function that he called entropy, noted S, and whose unit is the joule per kelvin (J/K). In this expression, Q is the quantity of heat received by a thermodynamic system, and T is its thermodynamic temperature.

Statistical thermodynamics then brings a new light on this physical quantity: it can be interpreted as the measure of the disorder degree of a system at the microscopic level. The higher is the entropy of the system, the lower its elements are ordered. They are linked together, capable of producing mechanical effects, and the greater the part of the energy that cannot be used to obtain work, that is, released incoherently. Ludwig Boltzmann expressed the statistical entropy as a function of the number Ω (Omega of microscopic states), or the number of configurations, defining the equilibrium state of a given system at the macroscopic level:

$$S = k * B \ln \Omega \tag{2.22}$$

where k {B} is Boltzmann's constant.

Pay attention that this new definition of entropy is not contradictory to that of Clausius. The two expressions simply result from two different points of view, depending on whether one considers the thermodynamic system at the macroscopic level or at the microscopic level [5].

2.6 Equilibrium

The concept of equilibrium is related to the stability of the system. It is very important in the study of phase diagrams. Different types of equilibrium can be defined (Fig. 2.5). The equilibrium is obtained when the total potential energy is at the minimum.

The chemical potential is linked in a particular way to the free enthalpy function, G, because it is about the only thermodynamic potential whose chemical potential is the partial molar quantity. According to Euler's theorem related to the first-order homogeneous functions, we can write for the free energy, which is an extensive quantity.

By considering the three situations, one can see that situation 1 is really a stable equilibrium. In this case, it is necessary to remember that the second law of thermodynamics states that an internal process may continue spontaneously as long as the dS change remains positive. The process will be stopped when dS $\leq$ 0 (1–22). This is the condition for equilibrium in a system.

By integrating dS, we may obtain a value of the total production of entropy made by the process. It has its maximum value at equilibrium. The maximum may be smooth, Δ S = 0, or sharp, Δ S < 0, but the possibility of that alternative will usually be neglected.

As an example of the first case, Fig. 2.5 shows a diagram for the formation of vacancies in a pure metal. The internal variable, generally denoted by ξ, is here the number of vacancies per mole of the metal.

As an example of the second case, Fig. 2.4 (taken from ref 20) shows a diagram for the solid-state reaction between two phases, graphite and Cr0.7C0.3, by which a new phase Cr0.6C0.4 is formed. The internal variable here represents the amount of Cr0.6C0.4. The curve only exists up to a point of maximum where one or both of the reactants have been consumed (in this case Cr0.7C0.3). From the point of maximum, the reaction can only go in the reverse direction and that would give Δ S < 0 which is not permitted for a spontaneous reaction. The sharp point of maximum thus represents a state of equilibrium. This case is often neglected and one usually treats equilibrium with the equality sign only, Δ S = 0.

If Δ S = 0, it is possible that the system is in a state of minimum ip S instead of maximum. By a small, finite change, the system could then be brought into a state where Δ S > 0 for a continued change. Such a system is thus at an unstable equilibrium. As a consequence, for a stable equilibrium we require that either Δ S < 0, or Δ S = 0 but then its second derivative must be negative.

It should be mentioned that instead of introducing the internal entropy production, Δ S, one has sometimes introduced dQ'/T where dQ' is called "uncompensated heat". It represents the extra heat, which must be added to the system if the same change of the system were accomplished by a reversible process. Under the actual, irreversible conditions, one has dS = dQ/T S. Under the hypothetical, reversible conditions, one has dS = (dQ + dQ')/T. Thus, dQ' = T Δ S. In the actual process, Δ S is produced without the system being compensated by such a heat flow from the surroundings.

If the reversible process could be carried out and the system thus received the extra heat dQ', as compared to the actual process, then the system must also have delivered the corresponding amount of work to the surroundings in view of the first law. Because of the irreversible nature of the process, this work will not be delivered, and that is why one sometimes talks about the "loss of work" in the actual process which is irreversible and produces some entropy instead of work, dW = dQ' = T Δ S.

A system is in a state of equilibrium if the driving forces for all possible internal processes are zero. Many kinds of internal processes can be imagined in various types of systems, but there is one class of internal process that should always be considered, the transfer of a quantity of an extensive variable from one part of the system, i.e. a subsystem, to another subsystem. In this section, we shall examine the equilibrium condition for such a process.

Let us first examine an internal process taking place in a system under constant values of the external extensive variables S and V, here, collectively denoted by

Xa, and let us not be concerned about the experimental difficulties encountered in performing such an experiment. We could then turn to the combined first and second laws in terms of dU, which is reduced as follows:

$$dU - Yud\xi a \quad Dd\xi = -Dd\xi \tag{2.23}$$

The driving force for the internal process will be

$$D = -(\partial U/\partial \xi)Xa \tag{2.24}$$

The process can occur spontaneously and proceed until U has reached a minimum. The state of minimum in U at constant S and V is thus a state of equilibrium.

$$dU = TdS - PdV$$

2.7 Definitions

In this section, we present all the definitions used in the book used for the description of thermodynamics in alloys Table 2.1.

2.8 Free Energy and the Free Energy Function

The free energy in a closed system is the energy that can be converted into work with respect to the state variables. Two different functions are created:

- Helmholtz free energy which refers to internal energy is expressed as

$$F = U - TS \tag{2.25}$$

the energy that can be converted into work at a constant temperature and volume.
- Gibbs free energy which refers to enthalpy and is expressed as

$$G = H - TS \tag{2.26}$$

the energy that can be converted into work at a constant temperature and pressure throughout a system. According to the fact that a change in the internal energy is obtained by variations in heat and work exchanged during the cycle, the functions are described.

Table 2.1 Expression of the different functions and variables

Electronic specific heat coefficient	Ce
Specific heat capacity at constant pressure	Cp
Specific heat capacity at constant volume	C_V
Magnetic component of the specific heat capacity	Cp^m
Debye specific heat function	Cp^L
Pressure	p
Quantity of heat	q
Work done in a closed system	w
Entropy	S
Absolute temperature	T
Debye temperature	T_D
Debye frequency	ω_D
Internal energy	$U(S, V)$
Enthalpy	$H(S, P) = U + PV$
Helmholtz Free Energy	F or A, $F(T, V) = U - TS$
Gibbs Free Energy	$G(T, P) = H - TS$
Chemical potential	$\mu i\,(T, P)$
Gibbs–Duhem equation	$\Sigma\, ni\, d\mu i = 0$

Note that one should make difference between differential quantities and small changes in the functions which are not state functions as q and w.

Enthalpy is defined as a function of Entropy and pressure at T, V constant.

In the same way, one has the two free energy functions described with respect to constant S, p, T and V.

Finally, the energy of the system is given by

$$H = H^{ref} + \int_{T_{ref}}^{T} C_p dT \tag{2.27}$$

Then we have the three relations giving the four state functions with respect to the different variables.

When the quantity of matter is changing, i.e. multicomponent systems, the Free Energy Function is a function of all variables as:

2.8.1 The Gibbs Energy Function for Pure Stoichiometric Substance

It is given by the relation which is a polynomial representation as a function of temperature and the coefficient are integers:

$$G = C_1 + C_2 T + C_3 T \ln T + C_4 T^2 + C_5 T^3 + \frac{C_6}{T} \qquad (2.28)$$

Gibbs energy is the central function of chemical thermodynamics. The expression in the case of a pure stoichiometric substance is in the form of the enthalpy of the formation of substance and entropy in the standard conditions (T = 298.15 K and Ptot = 1.bar) added to the temperature function of the heat capacity which is integrated over the defined range temperature:

$$S = S^{ref} + \int_{T_{ref}}^{T} \frac{C_p}{T} dT \qquad (2.29)$$

The Gibbs energy value is calculated from the Gibbs–Helmholtz relation G = H-TS.

One more thing concerns the definition of the Cp of a phase by going back to physics. Solid-state physics shows that the temperature dependence of the heat capacity C is explained by a quantum mechanical description of lattice vibrations (07Ein, 12Deb). Thus, one obtains the Debye function:

$$C = D\left(\frac{\Theta}{T}\right) \qquad (2.30)$$

where Θ is the Debye temperature which is a material-dependent constant. This description is theoretical and describes the heat capacity of a series of elements in their crystalline state very well.

As it is not necessary to refer to zero Kelvin as the reference temperature, a system of thermochemical data has been established on the basis of the standard element reference state (SER), taking place at room temperature (298.15 K) and 1 bar of total pressure (standard conditions). Conventionally, the value of enthalpy (H298) of the state of the elements which is stable under these conditions is set to zero. The entropy (S298) is given by its absolute value and the heat capacity at constant pressure, Cp, is described by a polynomial description (10):

$$C_p = c_1 + c_2 T + c_3 T^2 + \frac{c_4}{T^2} \qquad (2.31)$$

This equation describes the thermal properties of most substances within their experimental error limits and the parameters can be adjustable. In some cases, particularly when there exist polymorphic transformations it is necessary to split the temperature ranges in order to fit the functions within the experimental error limits.

From the standard Cp-polynomial and the known values of ΔH_{298} and S_{298}, one obtains the Gibbs energy equation as it was defined previously.

The coefficients Ci are now stored in a Gibbs energy databank such as SGTE. Understanding this expression can be done by assuming that the first two coefficients contain the contributions from both (ΔH_{298}, Cp) and (S_{298}, Cp) respectively, whereas the latter four are related to the coefficients of the standard Cp-equation. In this way all properties, especially the enthalpy and entropy values at room temperature, are the result of a calculation done from the coefficient tables in the standard compilations of, for example, JANAF or Barin, Knacke and Kubaschewski [10, 11].

When the temperature and enthalpy of transition and the coefficients of the Cp-equation of the phase at higher temperature are known the phase transitions of the first order are integrated into this data system, the G-function for the higher range is again derived from the integrals of enthalpy and entropy, but now based on the transition temperature instead of room temperature. Furthermore, the changes of enthalpy and entropy on phase transition need to be added.

The standardized treatment described above has also been used for substances which exhibit magnetic (second-order) phase transitions. To be able to handle the anomaly in the heat capacity that arises in such a case (see Fig. 2.2), it was thus far customary to split the temperature range around the Curie temperature into several small intervals such that the standard expression for Cp could be used. This procedure creates an unnecessarily large number of coefficients (e.g. eight times four Cp-parameters for Ni), and it also causes numerical difficulties because of the unusually large values of the parameters. SGTE has therefore adopted an approach suggested by Inden "which simplifies considerably the situation".

The magnetic contribution is treated separately thus leaving a well-behaved curve for the non-magnetic contribution to the heat capacity which can usually be described by one set of standard parameters for the entire temperature range. For the magnetic part of Cp, the critical temperature (Tc, either Curie or Neel temperature), the crystalline structure of the phase and the magnetic moment (ß) per atom in this particular structure are the prerequisites. One obtains for one mole of magnetic element:

$$G_{magnetic} = \mathrm{RT}\, f\left(\frac{\mathrm{T}}{\mathrm{T_c}}\right) \ln(\beta + 1) \tag{2.32}$$

Here, f is a structure-dependent function of temperature. It is different for the ranges above and below the critical temperature (for the details of this as used by SGTE, see [12]). Figures 2.1, 2.2, 2.3, 2.4 show the two distinct contributions from magnetism and lattice to the heat capacity and the resulting total curve.

A further additive contribution to the Gibbs energy, which is usually ignored because of its negligibly small value, stems from the pressure dependence of the molar volume. However, recent technical developments such as not only hot isostatic pressure but also more detailed research in geochemical phenomena have created a need for this extra contribution handled. SGTE has adopted the Murnaghan equation (for its mathematical description). This equation uses explicit expressions for molar volume at room temperature, $V°$, its thermal expansion, $\alpha(T)$, the compressibility at 1 bar, $K(T)$, and the pressure derivative of the bulk modulus, n (Bulk modulus = 1/compressibility):

$$G_{pressure} = V^0 \exp\left[\int_{298}^{t} \alpha(T)dT\right] \frac{[1 + nK(T)P]^{(1-\frac{1}{n})} - 1}{(n-1)K(T)} \qquad (2.33)$$

with $\alpha(T)$ and $K(T)$ polynomials of the temperature:

$$\alpha(T) = A_0 + A_1 T + A_2 T^2 + A_3 T^2 \qquad (2.34)$$

$$K(T) = K_0 + K_1 T + K_2 T^2 \qquad (2.35)$$

The necessary parameters have so far been assessed for a few substances, mainly metallic elements and some oxide phases of geological interest.

2.8.2 Mixtures, Solutions and Their Free Energy Functions

The solutions in metals and alloys as well as in some minerals are mostly defined in the same way. The notion of solid solution is a thermodynamic notion. It is a mixture of pure substances forming a homogeneous solid.

For a liquid, the notion of solution is quite intuitive: when a mixture is in solution, it cannot be distinguished in the components even under a microscope. Otherwise, it is precipitate or emulsion.

In solids, the case is a little different: in some cases, the solid is homogeneous on a macroscopic scale but the microscope makes it possible to distinguish different phases. The solid may have formed directly in this form (eutectic), but it can also have formed as a homogeneous crystal (crystallization), the initial solid solution then having undergone a demixing, generally during slow cooling.

Examples of solid solutions include salts and minerals which are crystallized from a liquid solution or a high-temperature solid phase.

When two or more component phases form a solid solution when gradually passing from one to the other by substitutions of their chemical elements, one speaks rather in this case of "isostructural series" or "isomorphic series".

Pure substances change from liquid to solid state without transition.

Intermediate materials are completely liquid at very high temperatures. The decrease in temperature is marked by the passage into a transition domain and the appearance of the first crystals. This domain is characterized by the liquidus and solidus curves of the phase diagram.

2.8.3   *Relative Molar Quantities*

The relative molar quantities are usually expressed by using the equations below.
Internal energy U(S, V).
Enthalpy H(S, P) = U + PV.
Helmholtz Free Energy F or A, F(T, V) = U − TS.
Gibbs Free Energy G(T, P) = H − TS.
Chemical potential μ_i (T, P).
Gibbs–Duhem equation $\Sigma\ n_i\ d\mu_i = 0$.
Activity $\mu_i - \mu_i^* = RT\ ln a_i$.

2.8.4   *Ideal Solutions*

According to Gibbs' theorem, a mixture of ideal gases is an ideal solution. This is included in the definition of ideal gases: in an ideal gas, the interactions between molecules are strictly identical, since they are zero. Thus, in a mixture of different chemical species in the state of ideal gases, all the interactions between the various species are equal to zero. Real gas mixtures which behave at low pressure like ideal gases are therefore ideal solutions, for example, air at atmospheric pressure.

Those mixtures are for example C5 to C8 linear paraffins: n-heptane and n-octane; mixtures of benzene, toluene and xylene; and mixtures of alcohols, for example, ethanol and propanol.

The behavior of a liquid chemical solution in which the solute is very dilute is close to that of an ideal solution. The colligative properties of this solution can then be determined by several laws demonstrated using the hypothesis of the ideal solution.

If an ideal liquid solution is in equilibrium with its ideal vapor, then the liquid–vapor equilibrium follows Raoult's law. The vapor (gas in equilibrium with liquid) is a mixture of ideal gases. On the contrary, by mixing 1 L of water with 1 L of ethanol a total volume of about 1.92 L is obtained. The ideal volume being 2 L, there is therefore a contraction of the mixture: the molecules of water and ethanol attract more strongly than the molecules of these pure liquids. The water–ethanol mixture is therefore not an ideal solution. Moreover, it also presents an azeotrope that Raoult's law cannot represent.

In the solid state, the copper–nickel (Cu–Ni) mixture can be considered as an ideal mixture, both in the liquid phase and in the solid phase. An ideal liquid–solid equilibrium follows the Schröder-van Laar equation generalized to ideal solutions (equivalent for liquid–solid equilibria of Raoult's law of L-V equilibria).

2.8.5 Real Solutions

Mixture functions

Xmix expresses the difference between an extensive thermodynamic quantity X of a real solution and the sum of the same extensive thermodynamic quantities Xid of pure substances i under the same conditions of quantity, pressure, temperature and phase as the ideal solution. It results as

$$X^{mix} = \sum_{i=1}^{N} n_i * R * ln \, x_i \qquad (2.36)$$

Note: do not confuse the quantity of the mixture X and mixing quantity Xmix.

2.9 Free Energy of Solutions

2.9.1 The Free Energy of Mixture of Phases

In heterogeneous equilibria, solids, liquids and gases are combined, and the thermodynamic quantities of the system are changing as a result of the mixing. The effect that mixing has on a solution's Gibbs energy, enthalpy and entropy.

A solution is obtained when two or more components mix homogeneously to form a single phase. It is called a homogeneous solution. Understanding the thermodynamic behavior of mixtures is integral to the study of any system involving either ideal or non-ideal solutions because it provides valuable information on the properties of the system. This allows combining our knowledge of ideal systems and solutions with standard state thermodynamics in order to derive a set of equations that quantitatively describe the effect that mixing has on a given phase thermodynamic quantities.

2.9.2 Gibbs Free Energy of Mixing

Unlike the extensive properties of a one-component system, which rely only on the amount of the system present, the extensive properties of a solution depend on its temperature, pressure and composition. This means that a mixture must be described in terms of the partial molar quantities of its components. The total Gibbs free energy of a two-component solution is given by the expression:

$$G^{mix} = \sum_{i=0}^{j} \left(\frac{\partial G}{\partial x_i} \right)_{p,T,Nj,\dots} \tag{2.37}$$

where G is the total Gibbs energy of the system.

The partial derivative is the partial molar Gibbs energy of component i, and it is the chemical potential of i in the phase:

$$\mu_i = \left(\frac{\partial G}{\partial x_i} \right)_{p,T,Nj,\dots} \tag{2.38}$$

$$\mu_i \leftrightarrow \overline{G}_i$$

with

And

μi is the chemical potential of the ieth component,

$\mu i°$ is the standard chemical potential of component i at 1 bar and.

pi is the partial pressure of component i.

The molar Gibbs energy of a mixture can be found using the equation:

$$G = \overline{G}_1 + \overline{G}_2 \tag{2.39}$$

The more simple way to understand is to take the standard molar Gibbs energy of a gas mixture at 1 bar, and P the pressure of the system. In a mixture of ideal gases, we find that the system's partial molar Gibbs energy is equivalent to its chemical potential which is the sum of the chemical potentials of component with respect to their concentration.

The Gibbs energy of mixing, $\Delta mixG$, can then be found by subtracting Ginitial from Gfinal.

This expression gives us the effect that mixing has on the Gibbs free energy of a solution. Since $\times 1$ and $\times 2$ are mole fractions, and therefore less than 1, we can conclude that $\Delta mixG$ will be a negative number. This is consistent with the idea that gases mix spontaneously at constant pressure and temperature.

2.9.3   Entropy of Mixing

Figure 2.6 shows that when two gases mix, it can really be seen as two gases expanding into twice their original volume. This greatly increases the number of available micro-states, and so we would therefore expect the entropy of the system to increase as well.

2.9.4 Enthalpy of Mixing

We know that in an ideal system the relation $\Delta G = \Delta H - T\Delta S$ is verified, but this equation can also be applied to the thermodynamics of mixing and solved for the enthalpy of mixing so that it reads.

This result makes sense when considering the system. The molecules of an ideal gas are spread out enough that they do not interact with one another when mixed, which implies that no heat is absorbed or produced and it results in a change as a function of the mole fraction so that.

of a solution will always be equal to zero (this is for the mixing of two ideal gasses).

2.10 Heterogeneous Equilibrium. Phase Equilibria; Equilibria Between Phases

First, it is necessary to give the definition of a phase such as a solid, liquid, gas and plasma. A phase is a physically distinctive form of matter. A phase of matter is characterized by having uniform chemical and physical properties. Then it is possible to define the notion of phase equilibrium. It was initially defined by J. W. Gibbs in 1898 [13, 14]. They were later taken up by Duhem and Van Laar [14]. It can be expressed as a closed system representing a finite quantity of material, separated from the outside by a wall. The wall can be traversed (in both directions) by flows represented by extensive quantities. The fundamental property of these extensive thermodynamic quantities is that they are "additive" over the whole system; these are state functions. The total volume of the system is the sum of the volumes of all its subsystems adjacent to each other: Energy, Quantity of matter (the number of moles of each constituent; i) and Entropy also. All the extensive quantities are positive, and their variations are algebraic. Clausius' relation defines heat exchange with the outside (see above) or gives entropy to the system.

Let's recall that one counts positively the exchanges of large quantities toward the system. Figure 2.2 represents a thermodynamic system in which we have separated small sets called subsystem σi (i = 1, 2, ..., n). Let S be the rest of the great system. The small subsystem σi permanently exchanges with S the quantities of heat, volume and matter. At thermodynamic equilibrium and over a finite time, the mean balance of these exchanges equals zero, which does not mean that a change takes place, but statistically it enters as much extensity of each species as it emerges. One can understand exchange is made by the existence of local fluctuations. The situation is much more complicated when different phases are present. A multiphase system is characterized by the simultaneous existence of several phases in the same given conditions, the two-phase system being the simplest case.

In general, equilibrium reactions in which the reactants and products are present in different phases are called heterogeneous equilibrium reactions; it is applied to the phase diagrams where the changes in the natures and compositions can be considered as chemical reactions. So we can extend this term to phase diagrams.

In chemistry of material study, it is often useful to differentiate the behaviors between global and local thermodynamic equilibrium. In thermodynamics, exchanges inside a system and with the outside are controlled by intensive parameters. For example, temperature controls heat exchange. The global thermodynamic equilibrium (ETG) means that these intensive parameters are homogeneous throughout the system, while the local thermodynamic equilibrium (ETL) means that these parameters can vary in space and time, but this variation is so great slower than for any point. This is why we can suppose that the thermodynamic equilibrium is always satisfied at the phase boundaries.

If the description of the system assumes very important variations of these intensive parameters, the assumptions made to define these intensive parameters are no longer valid and the system will not be in global equilibrium or in local equilibrium. For example, an atom needs to perform a number of collisions in order to achieve balance with its environment. If the average distance it has traveled moves it out of the boundary in which it attempted to enter equilibrium, it will never reach equilibrium, and there will be no ETL. This point is particularly important in the solid state where the structures are reconstructed. The temperature is, by definition, proportional to the average internal energy at the boundary in equilibrium. Since there is no longer such equilibrium, the mean temperature remains wrong.

2.10.1   The Phase Rule

Gibbs' phase rule was proposed by Josiah Willard Gibbs. This concept was defined in two papers [15, 16]. These papers are the basis of the problem of phase equilibria in multicomponent systems. As they were published in the journal "Academy of Sciences of Connecticut transactions", these papers were not so known in Europe. Fortunately a French scientist Pierre Duhem was in touch with him, and they exchange some correspondences, consequently the work of J. W. Gibbs diffuses into the whole of Europe. The first description of the phase rule is shown in those first papers. This rule is not really a rule but a theorem as it can be demonstrated. The so-called rule applies strictly to non-reactive multicomponent heterogeneous systems. Those systems should be considered at the thermodynamic equilibrium. Its expression is

$$F = C + 2 - \Phi \tag{2.40}$$

where F is the number of degrees of freedom, C is the number of components in the system (number of chemically independent constituents of the system, i.e. the minimum number of independent species necessary to define the composition of all phases of the system), Φ is the number of phases that are at thermodynamic equilibrium with each other and 2 represents the intensive variables T and p.

The number of degrees of freedom is the number of independent intensive variables that can be varied simultaneously and arbitrarily without affecting one another. An example of a one-component system is a system involving one pure chemical, while two-component systems, such as mixtures of water and ethanol, have two chemically independent components and so on. Typical phases are solids, liquids and gases.

Definition of a phase (Φ): A phase is a form of matter that is homogeneous in chemical composition and physical state. Typical phases are solid, liquid and gas. As an example, two immiscible liquids (or liquid mixtures with different compositions) separated by a distinct boundary are counted as two different phases, as it is observed in two immiscible solids.

The basis for the rule justification is that equilibrium between phases places a constraint on the intensive variables (13). More rigorously, since the phases are in thermodynamic equilibrium with each other, the chemical potentials of the phases must be equal. Then the number of equality relationships determines the number of degrees of freedom. For example, if the chemical potentials of a liquid and its vapor depend on temperature (T) and pressure (p), the equality of chemical potentials will mean that each of those variables will be dependent on the other. It means that mathematically, the equation μliq(T, p) = μvap(T, p), where μ is the chemical potential of a component, defines temperature as a function of pressure or pressure as a function of temperature.

The composition of each phase is determined by $C - 1$ intensive variables (such as mole fractions) in each phase. Then if we have Φ phases, the total number of variables is $(C - 1)\Phi + 2$, and number 2 represents the variables: temperature T and pressure p. As the chemical potential of each component must be equal in all phases, the number of constraints remains at $C(\Phi - 1)$. Finally, the number of degrees of freedom is obtained by subtracting the number of constraints from the number of variables and one obtains

$$F = (C - 1)\Phi + 2 - C(\Phi - 1) = C - \Phi + 2 \qquad (2.41)$$

This rule is valid in the case of the equilibrium between phases which depends on temperature, pressure and concentration, and it is not influenced by some other potentials such as gravitational, electrical or magnetic forces and by surface area, and in these last cases the rule must be adapted.

2.10.2   The Free Energy of Solutions

In normal language, a mixture is a dispersion of several substances which are more or less finely divided. In such a problem, one speaks of a heterogeneous or homogeneous mixture. This is depending on whether or not these substances are distinguishable. In Chemical Thermodynamics, the mixing action (often specifying mechanical mixing) is obtained only if the mixed substances form macroscopic domains (in the sense of thermodynamics) which are separated by a boundary, even if this division is not identifiable. When substances are mixed at the atomic or molecular level, it is not possible to constitute macroscopic systems.

Extensive state functions are additives such as.

Volume V Entropy S Internal energy U Enthalpy H Free Energy F Free Energy G	Molar functions are represented by $X = \sum_{i-1}^{n} ni\, X_i^{2*}$ where the X represent the functions' volume, entropy, internal energy, enthalpy, free energy function (Helmotz) and free energy function (Gibbs) of substance (40)

2.10.3 The Free Energy of Phases and Mixtures

An equilibrium is complete if its evolution is absolutely impossible under the conditions imposed on the system. When the temperature and pressure are kept uniform and constant at the boundary of the system and inside it, the Gibbs energy has reached an absolute minimum and can no longer decrease.

It can happen that equilibrium is constrained if certain types of evolution are impossible for kinetic reasons. For example, if the temperature and the pressure are kept uniform and constant at the boundary of the system, the free energy function reaches a local minimum and can no longer decrease as long as the stresses preventing certain chemical reactions or certain phase transitions (lack of sufficiently effective nucleation) are not lifted (for example, by the introduction of a minimal amount of a catalyst or seed crystals, or else spontaneously after a sufficiently long time). In either case, temperature and pressure are uniform inside the system, as well as its composition if it is single-phase or the composition of each phase if there is a number of phases.

At equilibrium, the state of a system is entirely described by its constitution (the number of moles noted I ni) and by two quantities which can be state functions or values imposed on the system (for example, the temperature T and pressure P, or temperature T and volume V). Its Gibbs free energy G, for example, can be thought of as a function of T from P and due to the simplicity of their differentials:

Certain state functions are expressed more naturally as a function of two particular quantities rather than others as is presented in Table 2.2.

All the state functions can be related to the state parameters as shown in Fig. 2.4.

Table 2.2 .

State functions	
Internal energy	U expressed in J (joule)
Enthalpy	H = U + pV expressed in J
Dilluu	Q imμνιιιιι iν J K⁻¹
Free energy (Helmholtz)	F = U − T.S expressed in J
Free energy (Gibbs)	G = H − T.S expressed in J

2.11 Variation of the Solubility with Temperature

In thermodynamics, mass solubility is a physical quantity denoted by the designation of the maximum mass concentration of a solute in a solvent, at a given temperature. The solution thus obtained is then saturated and the second phase precipitates. It is mainly the same in solids as in liquids. Likewise, molar solubility is the maximum molar concentration of the solute in the solvent at a given temperature. The mass solubility is expressed in g/L, and the molar solubility is expressed in mol/L.

The polarity of the elements as well as the geometric factors (coordination) influence the dissociation;

when the temperature of the solution changes, it can happen that, if the solute is solid, the solubility increases or decreases as a function of its enthalpy of dissolution;

if the solute is a liquid, the solubility increases with temperature;

if the solute is gaseous, the solubility decreases then the pressure increases;

if the solute is gaseous, the solubility increases.

Concerning the particular point of the solubility of gases in liquids, Henry's Law is strictly applied (22).

2.12 Conclusion

Finally, all the points developed in this chapter are resulting in four postulates.

Postulate 1 There are states of macroscopic equilibrium which are completely characterized by the specification of the internal energy U and the extensive parameters X0, X1, X1, …, Xn.

Postulate 2 There exists a function called entropy S. Entropy is a function of the extensive parameters and is defined for all the equilibrium states and it has the following property: the assumed values of the extensive parameters in the absence of external constraints are those which maximize the entropy on a variety of constrained equilibrium states.

Postulate 3 When Entropy of each constituent subsystem is a first-order homogeneous function of the extensive parameters, the entropy of the overall system is the sum of the entropies of the constituent subsystems. The entropy function is then continuous and differentiable. Obviously, it is a monotonically increasing function of energy.

Postulate 4 Entropy of any system disappears at the state for which

$$T \equiv \left(\frac{\partial U}{\partial S} \right)_{x_1, x_2, \dots} = 0$$

References

1. C.S. Barrett, T.B. Massalski, *Structure of Metals, Third Edition: Crystallographic Methods, Principles and Data (International Series on Materials Science and Technology)* (Pergamon, Oxford, 1981)
2. J.-O. Andersson, A. Fernandez-Guillermet, P. Gustafson, M. Hillert, B. Jansson, B. Jönsson, B. Sundman, J. Ågren, A new method of describing lattice stabilities. Calphad **11**(1), 93–98 (1987)
3. L. Kaufman, H. Bernstein, *Computer Calculation of Phase Diagrams with Special Reference to Refractory Metals* (Academic Press, New York, NY, 1970)
4. Z.-K. Liu, Y. Wang, *Computational Thermodynamics of Materials* (Cambridge, 2016)
5. M. Hillert, *Phase Equilibria, Phase Diagrams and Phase Transformations: Their Thermodynamic Basis* (Cambridge, 2007)
6. L. Lukas Hans, S.G. Fries, S. Bo, *Computational Thermodynamics, The Calphad Method*, (2007)
7. U. Kattner, H.L. Lukas, G. Petsow, B. Gather, E. Irle, R. Blachnik, *Z. fur MetallK* **79**, 32–40 (1988)
8. C. Giacovazzo, H.L. Monaco, D. Viterbo, F. Scodardi, G. Gilli, G.Zanotti, M. Catti, *Fundamentals of Crystallography* (Oxford University Press, 2011)
9. E. Mitscherlich, Premier Mémoire sur l'Identité de la forme cristalline chez plusieurs substances différentes, et sur le rapport de cette forme avec le nombre des atomes élémentaires dans les cristaux (in french). Annales de Chimie et de Physique, t. XIV **1820**, 172–190
10. C. Kittel, *Introduction to Solid State Physics*, 8th edn. (Wiley Ed, USA)
11. K. Kosuge, *Chemistry of Non-stoichiometric Compounds* (Oxford University Press, 1993)
12. P.A. Burrough, R.A. McDonnell, C.D. Lloyd, 8.11 Nearest neighbours: Thiessen (Dirichlet/Voroni) polygons. In *Principles of Geographical Information Systems* (Oxford University Press, 2015)
13. A.F. Wells, *Structural Inorganic Chemistry*, 5th ed. (Oxford University Press, 1984)
14. A. Guinier et al., Nomenclature of polytype structures. Nomenclature of polytype structures, report of the international union of crystallography ad-hoc committee on the nomenclature of disordered, modulated and polytype structures. Acta Cryst. A40, 399–340 (1984)
15. C.H.P. Lupis, *Chemical Thermodynamics of Materials* (Elsevier Science Publishing Co., 1983)
16. G. Inden, The role of magnetism in the calculation of phase diagrams. Physica **103B**, 82–100 (1981)

Chapter 3
Assessment of a Multicomponent System: The Ternary Space Model

3.1 Introduction

Beyond drawing, the thermodynamic aspect of the problem of the phase diagram is necessary in the modern study of multicomponent materials. At this stage of understanding, we will work on ternary systems, which are already a good initial description of multicomponent materials. High-order systems are more complicated in their evaluation procedure but, obviously, they result from the critical evaluation of ternaries. Assessment means that in the evaluation procedure, all data and the whole literature must be evaluated in the light of recent knowledge and theories. All the properties described the result of the phase rule enacted by Josiah W. Gibbs and the Gibbs–Duhem law which governs the phase relations, particularly at the boundaries of vicinal phases.

3.2 The Ternary Space Model

3.2.1 The Representation of Ternaries

The phase diagram (drawn from experiments or calculated by software such as ThermoCalc, Factsage and Pandat) must obey the laws of heterogeneous equilibria, and therefore the phase rule must be strictly observed. We should recall here that the phase rule governs the phase equilibria in a system [1]. In the solid state, it is a little difficult to A particular state of equilibrium is characterized by the nature and number of the participating phases to equilibrium and expressed as the following rule:

$$P + F = c + 2 \tag{3.1}$$

where P is the number of phases, c number of components and F degrees of freedom.

© The Author(s), under exclusive license to Springer Nature Switzerland AG 2024
J.-C. Tedenac, *Thermodynamics of Crystalline Materials*,
SpringerBriefs in Materials, https://doi.org/10.1007/978-3-030-99027-5_3

Fig. 3.1 Construction of
common tangent between
phases under equilibrium;
three phases in equilibrium
and two by two

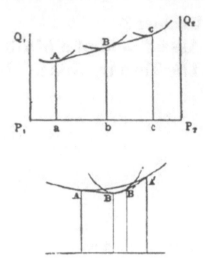

According to this definition, a good description of this problem was firstly presented by J. W. Gibbs in his book [2], and illustrated by Fig. 3.1 a and b which are the hard copies of the originals.

In this book, from a mathematical discussion all the examples of phase equilibrium in binaries and ternaries have been predicted, and we will see that the figure drawn by Gibbs is a predictive explanation of the isothermal representation of the equilibria. Let us recall here that J. W. Gibbs was the only mathematician explaining that this approach is general.

After this presentation, let us consider the ternary space model which can be built to understand the phase equilibria in ternary. Initially, such a system has 3 variables: Temperature, constituent content (A% B% C% if A B C are the three components). The lecture of such a spatial diagram is not very easy, and the description must be done by the way of bi-dimensional diagrams using only two variables and taking the third constant: Liquidus surfaces, secondary separation surfaces (solidus and/or melting surfaces), vertical sections and isothermal sections.

Fig. 3.2 shows the classical representations of such systems.

The description of a diagram presented in some books is done by a classification of phases taking part in equilibria, and it is a good nomenclature. It goes from 1 to 4 phases which are coexisting in the variable ranges. In order for a good understanding, let us consider the different situations where 1, 2, 3, 4 phases are present in equilibrium.

A—One phase is present.

In a ternary system where solid and liquid phases can be present and taking into account that pressure acts as a fixed parameter, the application of the phase rule shows that it is necessary to define temperature and concentration of components

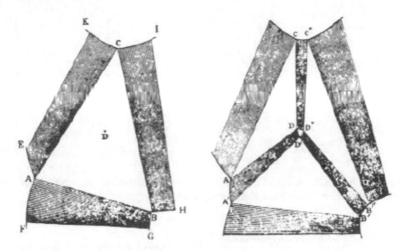

Fig. 3.2 Scheme of phase equilibrium in a eutectic system at the temperature and over it. These figures are taken from the original paper of J.W. Gibbs

in order to define the state of the system. The one-phase region is represented by a homogeneous three-dimensional space in the whole range.

B—Two coexisting phases.

When two phases coexist in a ternary equilibrium, two degrees of freedom exists, then and taking into account that pressure is fixed, two variables must be selected: Temperature and composition of one phase. The two-phase region is a space surrounded by two surfaces. These surfaces are connected by a set of tie-lines lying one surface to the other; they are straight lines and show the composition of the two equilibrium phases with respect to temperature. Due to the application of the phase rule, it is obvious that these tie-lines must be straight and give the relative amounts of phases with the ratio content of phase 1/content of phase 2.

C—Three phases coexist.

In such a case, one degree of freedom is present; the system is named monovariant. Then the choice consists of temperature or composition of one phase. The three-phase region is a volume formed by the three surfaces made by the sets of tie-lines. The intersections [3] are made by two other volumes of three-phase fields. Observing these equilibria, one must understand that at constant temperature the representation is a triangle as in Fig. 3.3b.

At a given temperature, the vertex is the tie-lines relaying the different two-phase fields.

D—Four phases coexist.

According to the phase rule with four phases at equilibrium, it results in an invariant equilibrium and in such a case these four phases are in equilibrium at

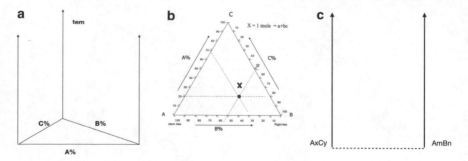

Fig. 3.3 a Spatial representation of a ternary. **b** Isothermal section and the representation of an alloy. **c** Isopleth section between two binary compounds or compositions

Table 3.1 The different reaction types in a ternary

Reaction type	Equilibrium between liquid and three solid phases
Eutectic [class 1]	Liquid $\rightleftarrows \alpha + \beta + \gamma$
Transitory reaction or "peritectic" [class 2]	Liquid $+ \alpha \rightleftarrows \beta + \gamma$ * this reaction is an intermediate between eutectic and peritectic. It is also sometimes named peritectic of class 2
Peritectic [class 3]	Liquid $+ \alpha + \beta \rightleftarrows \gamma$

one point as eutectic, peritectic and transitory reaction (or second type peritectic). All corresponding reactions are listed in the table. (Table 3.1).

3.2.2 The Situation with Compounds Existing in Binary Boundaries or Inside Ternary

The situation becomes a little more complicated in the systems in which appear binary compounds. Some simple rules must be followed in such a case. Then the system can be divided into subsystems where the phase rule is applied. One preliminary condition is needed for that: The tie-lines in the section must be in the plan of the section which means that all the compositions in this section belong to the quasi-binary section; as it is not obvious it should be experimentally proved. A single quasi-binary section divided the ternary into 2 parts, 2 quasi-binaries divided then the system into 3 parts and so on.

In the case of ternary compounds, the rules are the same but the resulting system becomes more complicated. So a simple rule appears: A ternary system is divided into n subsystems with n = b + 1, and b represents the number of binary intermediate phases. When binary and ternary phases occur in the same ternary, then it becomes n = b + 2t + 1, as b is the binary phase number, and t is the ternary phase number. The situation is shown in Fig. 3.4 a, b and c.

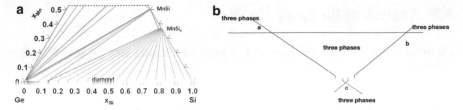

Fig. 3.4 **a** The tie lines showing the equilibrium between 2 phases. In this example equilibrium concern Si (C-diamond) and MnSi$_x$ in the system Mn-Ge-Si. **b** Equilibrium triangle of 3 phases separated by 2 phases regions. The boundaries of triangle are the primary tie-lines of the 2 phases fields

Generally, it is sometimes difficult to choose the stable phase fields existing in the system. An experimental procedure is used in the equilibrium calculations; it consists of the clear cross principle. When two sections are crossed, an alloy of composition x situated on intersections should be carefully studied and after reaching equilibrium the analysis of the samples indicate (composition of the mixture) and show which section is concerned, Fig. 3.5. This procedure can be also done by ab initio calculations. The calculations of the enthalpy of mixing should be done in the two sections step by step, and the crossing composition shows which end members are in equilibrium (then a tie-line must exist).

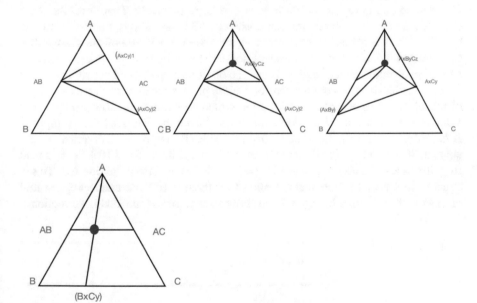

Fig. 3.5 Different possible cutting of a ternary system when commands are existing

3.3   Analysis of the Space Model

Different procedures can be done. We suggest an analysis step by step by using different sections.

3.3.1   Isothermal Sections

Firstly, we shall introduce the problem of matter quantities. Generally, it is accepted that the calculations have to be done in atomic percent, but sometimes and particularly in systems concerning steels and alloys, the calculations are expressed in mass percent. In any case, simple rules evidenced by Harman can be used to convert one into another. (Table 3.2).

In this book, the descriptions will be done in mole fractions easily convertible in % according to the given formula!

The isothermal sections are more suitable in the interpretation of a ternary system. Besides the liquidus projections (as below in Fig. 3.4), the isothermal ones are the second thing of importance.

At fixed temperature, an equilateral triangle is used to represent the composition of ternary alloys. Three vertices are named after the three components A, B and C. These three points represent 100% A, B and C, respectively. Therefore, the three sides of the triangle represent the compositions of 3 binary alloys. For each point on line BC, the side opposite to the vertex A has 0% A. Each side of the triangle has been subdivided into 10 parts by a set of points. The entire space is now divided into a set of small equilateral triangles. These can be further subdivided depending on the needs. Smaller divisions are needed for a more precise location of the composition of the alloy. Let us consider an alloy represented by point X as shown in the slide. What is its composition? How could this be read from the diagram? For example % A is read by measuring the distance from the side (BC) opposite the vertex A. It is equal to 30% (= CQ). Note that the lines are drawn at intervals of 10%. % B is read from the side AC. This is equal to 30% (= PX). %C is read from the side AB. This is equal to 40% (= AP). Note that the sum of the three is 100 or 1 and represents mol or mass in % or in unit. In Fig. 3.5, we show an example of an isothermal section.

Table 3.2 Conversion of mole% in mass%	$\dfrac{x}{100-x} = \dfrac{A}{B}\left(\dfrac{y}{100-y}\right)$ and
	$y = \dfrac{100\frac{x}{A}}{\frac{x}{A}+\frac{100yA}{B}}$ $x = \dfrac{100yA}{yA+(100-y)B}$
	where x % is the mass percent for one component and Y % expresses the mole percent of the same component

Fig. 3.6 An example of
isothermal section in the
ternary Zn-Sb–Te. It shows
the different domains: 2
phase fields with green
tie-lines and 3-phase fields in
red. Note that only one
binary compound exists in
this system: ZnTe

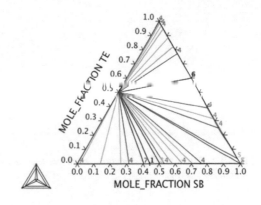

In Fig. 3.6, an isothermal section is shown; one can observe the trace of the equilibrium fields in the system Zn–S–Te for a section at 600 K. They are sharing, as single-phase fields, two-phase fields and three-phase fields as triangles. The border of the three-phase fields must be straight lines because they are the first tie-lines of the vicinity two-phase domains.

3.3.2 Vertical Sections and Isopleth Sections

To achieve a ternary analysis, we suggest using the vertical sections for the last step. For a person partially educated in phase diagrams, the vertical sections or/and isopleth sections are more difficult to analyze. However, these sections are a good tool for the interpretation of the microstructures and the crystallization sequences. We must make the divergence between these two representations; an isopleth section is a section where all the two-phase equilibria are in the plane of the section. The problem is that there is often confusion with isoconcentrate which is a section where one of the concentrations is constant. So an isopleth is a section with a constant concentration ratio, for example, an A/B = 0.3.

Considering a section at the isoconcentrate 0.56% C, it is only a section because if it was an isopleth the report would be explicitly specified such as B/C or C/B.

Moreover, it is sometimes possible to read in the publications the term quasi-binary, and this is obviously non-relevant. An isopleth section is a quasi-binary when there are only one-phase and two-phase fields with the condition: The tie-lines in the two-phase fields are contained in the composition plane. In Fig. 3.7, the section in between Te and the Sn, Pb alloy for 80% Pb is shown. It looks very different from a quasi-binary.

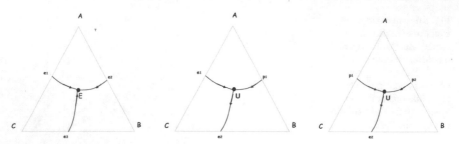

Fig. 3.7 Example of connections between the monovariant lines entailing ternary eutectic or transitory reactions

3.3.3   Liquidus Surfaces

It is obvious that the first step in the approach to the ternary system corresponds to the study of the liquidus surface and its projection. The first remark has to be made; these representations are interesting if the equilibrium between phases is respected and finally they can represent the correct situation. Drawn from the experiment or calculated diagram, results should be consistent with the concerned assessed binary systems in the studied ternary. A simple example consists of the phase changes for the same composition in all involved systems. This is the main point of the behavior. Second, it must be noted that the ternary invariant points should be related to the binary reactions (e or p).

The monovariant lines must be associated with the binary reactions and cannot meet within the ternary in a manner where the invariant ternary systems are connected in agreement with the phase rule. The main aspect is that four phases are connected only at invariant points (classes 1, 2 and 3 named as E, U and P).

These considerations are supported by the liquidus projection of the system as is shown in Fig. 3.7.

Usually, the assessment method used in ternary suffers from two kinds of problems: Insufficient data and/or sometimes older and insufficient knowledge of classical thermodynamics in the analysis of data obtained by calculation.

A full representation of the liquidus surface is given in Fig. 3.8 in the case of the Ni–Sn–Hf system.

3.4   Conclusion

As a conclusion, we will recall the general procedure which is used.

Phase diagrams are one of the most important sources of information concerning the behavior of materials, compounds and solutions.

They permit us to study and control important processes such as phase separation, solidification, sintering, purification, growth and doping of single crystals for

Fig. 3.8 An exemple of iquidus surface. This example is the ternary Ni-Hf-Sn, characterised by three ternary intermetallic aompaundn. I hù monovariant lines are the valleys and they are connected through invariant points. Isotherms are presented as level lines

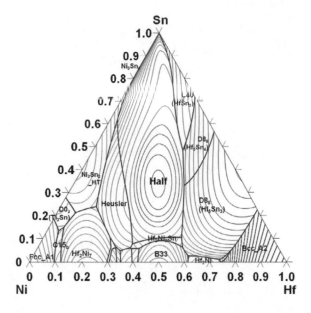

technological and other applications. They provide us with the knowledge of phase composition and phase stability as a function of temperature (T), pressure (P) and composition (C).

They can also assist in predicting phase relations, compositional changes and structures in systems not at equilibrium.

In such important applications, the description of systems and materials should be done carefully. This is the reason why the use of the CALPHAD methodology (done in a serious way) is the most important thing as one can describe systems or materials at equilibrium as well as systems out of equilibrium or near equilibrium. This allows showing that the description of the phases and materials leads to obtaining the ways for tailoring the materials. In fact, by calculating the thermodynamic functions of phases and systems, we can have all the information on these materials. Therefore, this information becomes of capital interest for alloy making, heat treatments or other things.

References

1. J. Willard Gibbs, Transactions of the connecticut academy of arts and sciences. Equilib. Heterog. Subst, (1876)
2. Donald E. Sands, Introduction to Crystallography, Dover publications Mineola, NY, 7 janvier (1994)
3. Marc De Graef et Michael E. McHenry , Structure of materials: an introduction to crystallography, diffraction and symmetry, 2nd edition, 8 octobre 2012, Cambridge University press, NY

Chapter 4
The CALPHAD Methodology: A Guide

4.1 Understanding Methodology

The modern approach to thermodynamics and phase diagrams is presently based on the CALPHAD method which is described in some modern books [1–3].

The CALPHAD (**CAL**culation of **PHA**se **D**iagram) method aims to describe the thermodynamic properties of a system by making consistent all existing data in a database. This procedure in done in three steps.

First is the collection of different data and their assessment. If the data are not numerous enough, it will be necessary to make some more and that is why the procedure is essentially a collaborative one (between physicians and chemists).

Second, in the light of these data, is the choice of the mathematical models of the Gibbs energy. Nowadays, a number of them have been developed in different parts of the world, and making a choice is more or less easy. This is necessary to implement for each phase of a system and the whole system at all temperatures and all compositions.

Third, the parameters of the mathematical models are optimized on the basis of existing data; often this can be done in the light of neighboring results. Obviously, these data are experimental or calculated (ab initio method) and fall under 3 main categories.

- Crystallographic data: they will be used in the description of the phase models and the description of variations in stoichiometry. Concerning phase modeling, it is necessary to take into account the coordination polyhedra, the nature of atoms or ions and the defects (type and crystallographic sites where they are).
- Thermodynamics: it is mainly not only the data of enthalpy of formation and the enthalpy of mixing in the liquid phase but also all the data deriving from Gibbs energy (electromotive force measurements, chemical activity, partial pressure, C_P, etc.).
- Diagrammatic: all data of phase diagrams such as liquidus, invariant temperatures and domain of phase composition is measured by thermal analysis.

© The Author(s), under exclusive license to Springer Nature Switzerland AG 2024　　　57
J.-C. Tedenac, *Thermodynamics of Crystalline Materials*,
SpringerBriefs in Materials, https://doi.org/10.1007/978-3-030-99027-5_4

Then one can seat the modeling of the system. Two ways are possible: using an optimization module (as PARROT in Thermo-Clc) which is used in all existing calculation software. The objective of an optimization module is to provide parameters for use in thermodynamic and thermophysical models. This is done by fitting the parameters of the model with a large number of experimental observations of magnitudes describing equilibrium states or dynamic processes in heterogeneous systems with several components. This type of module allows the user to interactively modify phase descriptions, model connections, basic thermodynamic parameters and so on.

The calculation module (which is named "POLY" in ThermoCalc software) is a phase equilibrium calculation module. This module specifies the conditions of calculation and therefore of use. It is obviously necessary to start the calculations in a field with two phases in equilibrium then the mapping does the rest. When the modeling is ended, one obtains a database containing the description of the Gibbs energies of each phase of the system versus temperature and pressure. At the end, a submodule handles the processing and the plot of calculations by using two by two the free parameters and functions of the system. This base is then sufficient to re-draw all the thermodynamic data as well as the phase diagram which is, ultimately, a representation of the minima of the Gibbs energies as illustrated in the following diagrams. Two examples are chosen for the understanding. First, we choose a system as Cu–Ni with 2 components showing a quasi-regular solution. In such a case, the phase diagram looks very simple (Fig. 1). When the constituents are very different from a constitutional point of view, the diagrams are very different. One can see an example in Al–Zn, Fig. 2; this diagram is well-known and shows significant differences in the diagrammatic approach. In this case, it is interesting to consider the different $G_m{}^\varphi$ curves which help to understand the behavior of the system. In this example there are two specific things:

(i)   a miscibility gap exists in the solid state and, as is known; (ii) two different phase compositions of the same solid solution $(Al,Zn)_1$ and $(Al,Zn)_2$ are identified by two points 1 and 2 ($\times 1$ l–l $\times 2$) in the phase separation curve at constant temperature (Fig. 9). Then they collapse at the maximum at zero tangent ($x_1 \equiv x_2$); in the Gibbs energy diagram, one can see that the lowest curve shows two minima. This problem is represented at the equilibrium between $(Al,Zn)_1$ and $(Al,Zn)_2$ with a change in the curvature at the binodal composition. As the temperature grows, the two minima approach and join at the maximum temperature which is called "critical temperature". These two minima correspond to the compositions of the two equilibrium phases at the considered isotherm temperature. Then concerning a phase in equilibrium with a constituent as Zn, for example $(Al,Zn)_2 + (Zn)$, a common tangent show also clearly this fact. In this case the invariant is named monotectoid plateau.

Another thing concerns the problem of the metastability of phases. This is a pregnant problem in the application of industrial materials. The case of the Cu–Mg phase diagram explains the different possibilities to study the phase stabilities. This system contains two intermetallic phases (one stoichiometric and one slightly non-stoichiometric). An accurate calculation of the different possible equilibria (taking into account the phases as stable or metastable, one by one or both) between the boundary phases and the two intermetallic compounds is presented in Fig. 3 ; the dotted lines show these different possible phase diagrams when 1 or 2 intermetallic are considered as metastable. In a software such as ThermoCalc, it is presented as "suspended phases". Such a problem is easily transferable to the ternaries and high-order systems, but it needs that the binaries must be very well assessed.

As told in the introduction, implementation of the CALPHAD method by several software has been developed (for example, ThermoCalc, Pandat, FactSage and Open-Calphad). One should choose a software for doing calculations. We don't give indications on the choice but all are built by using very closed algorithms and in this tutorial, we have chosen to present cases treated using ThermoCalc and sometimes with Pandat.

Moreover during the development of this procedure, different interesting concepts were developed and need to be presented for a better understanding of the method.

4.2 The Phase Model

The phases (solid, liquid or gas) are systematically described by a sub-lattice model containing one or more sub-lattices. By considering such designation of the phases, $(A,B)x-(C)y$, the model contains two sub-lattices. What is the concept of sub-lattice? The concept of sub-lattice consists in the arrangement of all atoms in part of the main lattice; that is why it is called sub-lattice; the sum of all sub-lattices is the real lattice. In each of sub-lattice, it is possible to place one or more atoms, as well as vacancies (denoted by V_A). This is an example of 2 sub-lattice cells: Caesium chloride, CsCl, is written as $(Cs, Rb, V_A)-(Cl, V_A)$ showing possible exchange of cations on the same type and V_A are used on the Cl site. It gives one sub-lattice of cations and one sub-lattice of anions. The vacancies are introduced in order to take into account the possible departure to stoichiometry in the compounds.

For 10–20 years, complex non-stoichiometric phases have been modeled according to their representation of their chemical and physical reality.

They will be modeled by taking into account that the major problem concerns the correct use of the position of vacancies or anti-site defects in order to describe the reality of the deviation to the stoichiometry. A sub-lattice will be used for each crystallographic site (constituting the sub-lattices of the structure). But in certain cases, we can simplify the description of the phase as much as possible by associating together "similar" sites, which implies a good knowledge of the crystallographic structures (and defect properties in the structure). This explains Chap. 1 of this book.

When it is necessary to build a database containing more than two elements, obviously homogenization of the models is needed. The same description (with the same number of sub-lattices) for each phase is required (for example, the liquid or two phases of the same structure in 2 different binaries, even if there is no solubility).

This approach is commonly used in the liquid phases. Thus, there are 4 models to describe a liquid.

The classic model: All the elements are on the same sub-lattice (A, B, C,…)1; it is the case where very few interactions between components exist.

The associated model: It is the most widespread in the liquid phases. It is used when significant interactions exist in the liquid (leading to sharp mixing enthalpy curve and/or offset by 50/50%); it will be assumed that a local order is established in the liquid with a set of AxBy "molecules". A single sub-lattice such as (A, B, AxBy)1 is therefore used. Due to the weak interactions in the intermetallic compounds, this model is not so used in this case.

The ionic model: In such a model, the liquid must have a clearly ionic character and in this case, the liquid is described by adding several sub-lattices, taking into account the addition of ionic elements or vacancies. This is clearly used in ionic ceramics.

The quasi-chemical model: It was developed particularly by the Montreal school of thermodynamics working with FactSage and is used mainly by some groups around the world. The advantage concerns the ionic systems; they describe that this model takes better account of interactions around pure elements. The main problem is that this model is not compatible with other models. Therefore, to describe a ternary system, one would have to have the 3 binary liquids described with this model, which is more complicated.

The mathematical model used for liquid and solid solutions is based on a description using a mathematical model: "the Redlich–Kister model". Then the Gibbs energy of each phase is described by the equation:

$$G^{EX}/(R \times T) = \Sigma a_i \times x_1 \times x_2 \times (x_1 - x_2)_i \qquad (4.1)$$

where the summation is from $i = 0$ to n terms.

In the case of stoichiometric intermetallic phases (or those where the homogeneity domain can be considered as negligible), also called line compound, if high-temperature Cp data aren't reported, the Gibbs energy is referred to the constituent elements in their reference state using an expression such as that described by the Kopp–Neumann and Muggianu models [4] which are by far the most used.

The reference terms of each element and component (in solid as in liquid phases) are given in the SGTE database (Scientific Group Thermodata Europe) whose first version was published in 1991 by Dinsdale [5]. Several updates are available, the most recent (2010) being implemented in ThermoCalc [6] as the PURE5 database.

In the process of assessment, the challenge is to optimize the values of different parameters a and b in order to get the best representation of all available data.

When three elements are present in the system (for example in ternary liquids), we can use.

1 single parameter ^0L assuming a symmetrical interaction whose minimum is centered on 33%/33%/33%;

3 parameters, under these conditions, ^0L corresponds to the interactions between the first element (placed in alphabetical order) with the other two, ^1L between the second and the other two and ^2L between the 3rd and the others, etc.

Different types of files are needed for the calculations.

CALPHAD approach is a thermodynamic analysis of systems and not only the description of phase diagrams. CALPHAD describes the phase equilibria, with a realistic combination of the principles of basic thermodynamics, and mathematical formulations of thermodynamic functions.

Finally, as surprising as it is, the CALPHAD approach taken its roots in the research which was born in 1908 from the work of Van Laar. He was the first to apply the energetics concepts of Gibbs' phase equilibria theory. At that time, he didn't have the digital input required to convert algebraic expressions into a phase diagram, and the concept was set apart. Finally, during the 1970–1980 modeling, the phase diagram was developed by using computational algorithms, and this was the first time when CALPHAD was established and recognized as a tool in material modeling. A high level of empiricism was present at first but gradually decreases with time and with the use of new concepts, new software and more available experiments. This approach can be described by a sentence written on the site calphad.org (Introduction) which really leads to understand the development of this approach:

> We believe that substantial progress be made in a short time if we could arrange to work together for a week at one of our facilities to define problems, dissolve, carry out certain individual activities, and meet again for a week at a second facility to compare the results and the table of future activities.
>
> by Larry Kaufman and Ibrahim Ansara 1973 [7].

Then the procedure was born and led to performing more and more calculations. The CALPHAD procedure of calculations is briefly described in Fig. 4 . First of all, this scheme is showing roughly the assessment procedure. It will be useful for understanding the process.

Then the applications of this concept can be applied in an algorithmic way using software.

4.3 Understanding the Software: The Software Architecture

The text below refers to ThermoCalc but the procedure is very similar to other software. ThermoCalc (TC in the following text) can be used in two ways: the control console and the graphical interface. In this part, we will mainly discuss the command's part. The graphical interface is generally well developed and is sufficiently "instinctive" from the moment it is better to use the master command mode. We switch from one mode to another via the "Switch to Console Mode" button.

Starting the software is made through the system root level (SYS), and it is possible to navigate from one module to another with the command "go", for example writing "go poly or go p" to go to the POLY module and do calculations. Returning to a former level below is made by using the "back" command. An important thing is that the memory of previous calculations is taken into account in the next.

One useful command is "?". It lists all the commands available in a module at the stage of calculations. Some of the modules include.

DATA to manage databases;

GIBBS to modify the TDB file (addition of phases, interaction parameters);

POLY to calculate an equilibrium;

PARROT to optimize a system (we will not use it now because it takes a level of software knowledge);

SCHEIL is useful to make solidification calculations;

There are other modules which can be used for specific problems (BINARY, TERNARY, SCHEIL, etc.).

Each command can be used as a whole or replaced by a shortcut. But there should be no ambiguity. The commands and shortcuts are given in Table 1. For example, for the command "set-conditions", we can use "s-c" or "set-c" or "set-cond". A contrario the two commands "list-equilibrium" and "list-conditions" exist, and it should be precise. It should be used at least "s-e" and "s-c". Command lines are not case-sensitive. One more thing: depending on the TC version those commands can slightly change.

Before making the first calculations on the ternary, it is necessary to have a database containing the description of the phases of each of the 3 binaries and a rough description of the ternary compounds which are inside. We will start from the principle that the system has been modeled and/or that the parameters have been published in the literature. In recent articles (especially in the Calphad journal), TDB files can be provided by the author as "supplementary materials". When having problems with writing the TDB, asking the author is nevertheless a good way.

4.4   Database and Calculation Modules

The ThermoCalc software uses and generates several types of files. One is necessary to build the TDB which summarizes the whole database and the basis for calculating functions and diagrams.

All files used in the software are as follows.

– TDB: this is the file containing the complete database of a system. This is the file that is usually provided at the end of the calculations. It allows drawing all the functions Gi, Hi, Si, T, p, Xi,.etc. The file can be edited and checked line by line. This file can be used in ThermoCalc and Pandat and others by a few syntactic adaptations.

– TCM: these commands are macro commands. They are made in order to not rewrite command lines several times in the same calculation, and this accelerates the calculation process. It contains all command lines which will be executed (by double-clicking directly on this file). The file can be edited and modified as much as we want.

- EXP: this file is very important. It contains all the data (experimental or/and calculated). It is also used to paste the experimental data on the graph (for example, ATD measurements on a former published phase diagram). The file can be edited.
- POLY: it is generated by the software during a calculation. It is used during calculations, and it is not possible to edit.
- PAR: it is a file created during a system optimization (PARROT module). It is impossible to edit this file.
- POP: it is a file which contains all the thermo data which will be used to optimize a system. It partially contains the EXP file. This file can be edited too.

4.5 Creating the TDB File

The first thing to do is to recover the data relating to the pure elements. This information is on the top of the TDB module.

The TDB files are used for calculations in all ThermoCalc files. Certain lines are starting with $ (dollar); in such cases they are inactive. Nevertheless, they are useful for understanding the thread of the file and can be used to put comments or temporarily suspend one or more parameters during a calculation.

Let's analyze these files step by step. A TDB file contains the following:

A—General data on the elements. These data are necessary to put in the equations for the calculation(example: ELEMENT AL FCC_A1 2.6982E+01 4.5773E+03 2.8322E+01!). In such order it means that the line concerns the element Al and, under standard conditions, it is in the structural form FCC_A1 (name of the phase) of atomic mass 26.982 g/mol; it results in H298-H0 with 4577.3 J/mol and S298 with 28.322 J/mol/K.

B—All the functions will be used afterwards. It is not necessary to use all these functions, but often it simplifies the calculations. In general, the functions for the elements are present in each state: GHSERSI (the Gibbs energy of the SI in its reference state), GLIQSI (the Gibbs energy of the liquid Si), GBCCSI (the Gibbs energy of the element in the form BCC), GFCCSI (the Gibbs energy of Si in the form FCC), etc. Each function is described in the form of "FUNCTION GBCCSI 298.15 + GHSERSI + 47,000-22.5 * T; 3600 N!". This means that the GBCCSI function is GHSERSI + 47,000-22.5 * T in the temperature range 298.15–3600 K, and there is only one temperature range (N). Concerning the G function of the liquid or the elements in their reference state, there are several temperature ranges (Y instead of N) as described in the literature. They represent the different fitting of functions versus temperature. Finally, the (!) means that the command line is finished.

C—Description of the phases. It is needed to completely describe the phases of the system (gas, liquid and solids). For these phases, firstly one should declare the phases one by one in a format as this: (PHASE LIQUID% 1 1.0!). This means that a LIQUID phase is created. This phase contains a single sub-lattice with 1 as stoichiometry. The second line describes the elements which are present in the phase. As an example in the well-known Al–Si system; the components of the LIQUID phase are AL and SI on the first sub-lattice (placed between::) it results in(CONSTITUENT LIQUID: AL, SI:!). Then the description (in sub-lattice) of the liquid phase is obtained: (Al, Si) 1. The following lines correspond to the parameters of each constituent in this phase. Let us understand this building of lines.

(PARAMETER G (LIQUID, AL; 0) 298.15 + GLIQAL; 2900 N REF2!). It means that the Gibbs energy function, parameter G of LIQUID with only Al (therefore the G of pure liquid Al), between 298.15 and 2900 K is described by the GLIQAL function. The reference of this parameter is named REF2 and it is put before the !.

The list of references is in the form (REF2 'PURE3 - SGTE […].' REF3 'PURE1 - SGTE […]'!). In fact each reference is given freely between "at the end of the references!" Should be added then it will be possible to add references in the same format.

D—Adding interaction parameters.

As now a database (TDB) file has been created, this file currently only contains element data. It is necessary to add the binary interaction parameters that we will have retrieved from the publications and the data concerning the compounds. To do this, one should simply add the command lines in the TDB.

These things are summarized below for the example of the Al–Si system:
Open ThermoCalc and switch to the Console Mode if necessary.
This creation is made through the command lines shown below.

go data (go da)	Go to the data module
switch (sw)	Change database
PURE	The one that contains the pure elements. We will choose USER to make a TDB
def-el AL SI	Defines the elements of TDB
get	Data extraction for calculation
go gibbs	Go to the Gibbs module
l-data	List the database
file	It can be listed on the screen (SCREEN) or put into a file (FILE)

The TDB file is created in the desired folder. The file here will be called «*,**i.TDB.

N   Option N to create a database in "USER" format.

One will have to use this procedure (the first 5 lines) each time a database is opened; we will see that when writing the macros.

This produces an editable TDB file.

A full example is shown as Ca–In. In the system Ca–In (described in reference 15), we find liquid equation in Table 2. The critical evaluation of the phase diagram and the thermodynamic data available in the literature allowed the thermodynamic optimization of the Ca–In system.

The CALPHAD approach methodology (CALculation of PHAse Diagram) could be used by these authors (99).

Intermetallic compounds (i.e. Ca8In3, CaIn2, Ca2Sb, Ca5Sb3, Ca11Sb10 and CaSb2) have been described as stoichiometric phases, in the absence of further information on possible non-stoichiometry. The order–disorder transition between Φ BCC_B2 (i. um) and BCC_A2 (0-x a) was considered and a single Gibbs free energy function to describe both ordered and disordered parts. By means of ab initio calculations, the enthalpies of the formation of intermetallic compounds at 0 K were calculated and used in thermodynamic modeling. A set of self-consistent thermodynamic parameters for the Ca–In system was finally obtained.

They are the interaction parameters in the liquid and models for the solid one.

In such systems, the liquid model is already quite elaborated, since there are interaction parameters of three order (0, 1 and 2), and each parameter has a temperature variation.

We will therefore add the following line in the TDB. This line is put before those describing the solid phases (FCC_A1, BCC_A2, BCC_B2, etc.).

Ca–In Liquid model (CaIn)1   Parameters: ^{0}L Liquid = −148,791. 6 + 23. 01 T ^{1}L Liquid = 15.705. 17 − 18. 0 T. ^{2}L Liquid = 36,036. 0.

In this example the authors are using 3 parameters. In most cases only 2 are necessary.

Of course, for a better understanding, we will add the reference (in the list) in the form ($ Song Qin et al.: CALPHAD: Computer Coupling of Phase Diagrams and Thermochemistry 48 (2015) 35–42).

One proceeds in the same way with each phase described in the paper. One should pay attention to certain phases such as BCC_A2 which are described with two sublattices, the second being composed of vacancies. When writing sub-lattices, it needs separation by:

Declaration of binary/ternary phases.

Most of the time, adding new phases is necessary as they were discovered by solid-state examinations. For a stoichiometric phase, we have to choose between line compounds and compounds with a slight departure from stoichiometry. We take the example of a SiB3 or SiB6 phase in the binary B–Si (Lim publication) where the two kinds of phases are present.

Don't miss that stoichiometry corresponding to the compound must be correct (!). Obviously, the matter quantity must be equal to one mole of compound. If it cannot be the case (for some conveniency), the matter quantity has to be an integer.

It is the simple case where, since we have no Cp data measured at T > 298 K, we use a simple model referring to the elements in their standard reference state. At this time, G (AxBy) is written as ΔfH298K − ΔfS298K + xG(A) + yG(B). In such a case, the SiB3 phase is very simply described by 2 sub-lattices: one with Si and the other with B, in the 1/3 stoichiometric ratio. We will be able to write our sub-lattice in two forms: (Si) 0.25 (B) 0.75 or (Si) 1 (B) 3. If the second form is perhaps more instinctive, the first has the advantage of being related to one mole, which I personally prefer (we can thus directly compare the enthalpies of formation from one phase to another). We will write in this case: one with Si and the other with B, in the 1/3 stoichiometric ratio. We will be able to write our sub-lattice in two forms: (Si) 1 (B) 3 or (Si) 0.25 (B) 0.75. If the first is perhaps more instinctive, the second has the advantage of being related to a mole, which I personally prefer (we can thus directly compare the enthalpies of formation from one phase to another).

In the case of a phase with a variation in stoichiometry, the problem is slightly different. Let us take the case of Al3Nb which was published by Witusiewicz et al. in an article concerning the Al–Nb–Ti ternary system, only a part of the data is taken into account. The phase is described with two sub-lattices. The authors assume that the solid solution is formed by the substitution of Nb at the Al site and Al at the Nb site. Sub-lattices are therefore (Al, Nb)3 (Al, Nb)1, highlighting the majority species with the formula Al3Nb. In this case, we will have to define, ideally, the Gibbs energy of all the phases generated by the combination of the 2 elements on each sub-lattice (Al: Nb; Al: Al; Nb: Al and Nb: Nb). So we will write the lines as in Table 3.

This example can be extended to the related ternary phase written as AL3M (M: Nb, Ti in variable content). This table is taken from this paper and explains this fact. Obviously, understanding such a system leads to be able to manage all the systems.

The article of Witusiewicz is about an Al–Nb–Ti ternary system, so we only consider part of the data. The phase is described with two sub-lattices. The author assumes that the solid solution is formed by substitution of Nb at the Al site and Al at the Nb site. Our sub-lattices are therefore: (Al, Nb)3 (Al, Nb)1. The majority species having the formula Al3Nb are underlined. In this case, we will have to define, ideally, the Gibbs energy of all the phases generated by the combination of the 2 elements on each sub-lattice (Al: Nb; Al: Al; Nb: Al and Nb: Nb). So we have Table 4.

By convention, when adding parameters, we choose G for the description of a Gibbs energy of an end-member and we choose L for the interaction parameters. A priori, we must be able to write G everywhere without this changing the results, but it is better to remain consistent.

4.6   Basic Calculations with TDB

Once a TDB database is written, various calculations can be performed. Some examples which will be used in the following will be shown.

For these calculations, we will use the POLY module. First of all, we must start in ThermoCalc by selecting our database as well as the elements. For that, we use the commands already mentioned (Table 5).

Equilibrium calculation

This is the basic calculation. Starting calculations need to fix the setting conditions, such as the total mol number (N), the composition (x(Si)), the temperature (T) and the pressure (P). It is also possible to impose a number of coexisting phases.

Then during the calculations, it is possible to give several statuses to the phases via the command (ch-st), change-status.

The phases can be:

ENTERED: the phase contributes to equilibrium (by default).

FIX: phases are imposed.

SLEEPING: the phase does not participate in the equilibrium, but its Gibbs energy is calculated.

SUSPEND: the phase is excluded from calculation.

For example, let's determine the phases in equilibrium in the Al–Si system at 60% Si and 800 K under standard conditions. The commands are given below:

s-c x (Si) = 0.6 We define the composition: set-condition; x (Si) is the atomic fraction of Si where the calculations start.

s-c T = 800 N = 1 P = 1E5 Then come the variable conditions. It is possible to put several conditions in a single line.

c-e compute-equilibrium.

list-e list-equilibrium The listing of equilibrium conditions is shown in order to correct some problems.

VXCS is a combination of options, to change you can type? to see the others. Then one can obtain on the screen: the conditions of the calculation, the elements present, the list of stable phases under these conditions, for each phase, the percentage of it as well as the percentage of each of the elements in the phase.

Note that if a mistake is made, removing a condition is simple; just write, for example, "s-c T = none". By default, T is in K.

Thermodynamic data (enthalpy formation, Gibbs energy, etc.)

After calculating an equilibrium, you can display any thermodynamic data with the "show" command. Of course, first it is necessary to define the reference states (set-ref-state). For example, we can use.

show AC (AL) for the chemical activity of Al with the default reference states.

ACR (AL) show for the chemical activity of Al with the reference states defined here.

show MU (AL) for the chemical potential of Al with the default reference states.

show MUR (AL) for the chemical potential of Al with the reference states which are defined here.

The other quantities are for example:

HM: enthalpy of the system.

HM (BCC): enthalpy of the BCC phase.

SM and SM (BCC): same as entropy.

GM and GM (BCC): same as Gibbs energy.

AC (AL): chemical activity of aluminum.

AC (Al, GAS): chemical activity of Al in the gas phase.

DGM (BCC): the driving force of the phase (0 if the phase is stable, positive otherwise).

...

Note that all the derived and partial quantities can be obtained by differentiating. For example, the derivative of HM (BCC) with respect to temperature is written with a dot: "HM (BCC).T".

Other things.

In general, ThermoCalc is not case-sensitive, but it is better to prefer to write the atoms (or the vacancies) in uppercase (AL, SI, NB, VA, etc.); the software sometimes do not recognize the elements when they are written in lowercase.

Transcription of a TDB (from an advertisement).

Before doing the first calculations on the ternary, it is necessary, of course, to have a database containing the description of the phases of each of the 3 binaries. It is assumed that the system has been modeled and that the parameters have been published in the literature. In recent articles (especially in the Calphad journal), TDB files may be provided by the author as "supplementary materials". If we really have trouble writing the TDB, it is possible to ask the authors.

Writing a database.

Creating the TDB file

The first thing to do is to recover the data relating to the pure elements. For this, we use the TDB module.

Open ThermoCalc and switch to Console Mode if necessary. The the command lines are presented in Table 6.

Graphs drawing

When having a well-done TDB and knowing how to do simple calculations, all kinds of graphical representations can be obtained. In this part, it is mainly interesting to use macro instructions (named TCM file) which can easily be duplicated and adapted to each system.

A macro is a set of commands that are bundled together in a {.TCM} file. Opening them is easily done by using, in the console mode, the "mac" command or opening them by double-clicking directly.

At the end of the macro, the software closes automatically the calculations. To avoid this, we will add the "set-inter" command at the end of the macro and continue more calculations.

For any calculation via a macro, the main steps are.

Choice of TDB and elements (already described);

Calculation of a starting point (already described);

Extension of calculation to a domain of compositions/temperatures;

Plot of the diagram.

The extension of a calculation is done in the POLY module. After the calculation of the starting point, we will define the axes and the fields of calculations with the command "set-axis-variable". We use for example:

s-a-v 1 x(Al) 0 1 0.01	on the x-axis, the content x(Al) vary from 0 to 1 with a step of 0.01
s a v 2 T 10 3000 10	on the y axis, temperature (T) vary from 10 à 3000 a with a step of 10

Please note: the calculation can ONLY be extended to state variables that have been used to define the starting point. For example, if we wrote "s-c x (Y) T = 500 n = 1 p = 1E5", we cannot write "s-a-v 1 x (X) 0 1 0.01".

Of course, as smaller the calculation step is, the sharper the graph will be. On the other hand, the computation time will be considerably long (example: up to several minutes for an isopleth section in a ternary).

To extend the calculation, there are two possibilities depending on what plot is needed. If, for example, drawing a binary diagram or an isothermal section in the ternary is necessary, sweeping x1/x2 or x1/T space in both dimensions is required. On the other hand, plotting at fixed T the evolution of the enthalpy of mixing in the liquid, sweeping only needed on an axis × 1. In the first case, we will therefore use the "map" command after having defined two axes. In the second case, we will use the "step" command after having defined a single axis.

Depending on the versions of ThermoCalc, it is mandatory to add command lines in order to extend the calculations in the positive and negative directions (add the commands "add 1", "add −1" for axis 1 and "add 2", "add −2" for axis 2).

Once this part is finished, go to the post-processing module done via the "post" command. Define which representation is needed via the "set-definition-axis" command, Table 7.

Finally, the graphic representation can be improved, by changing the title of axes, for example, or by changing the limit's definitions of the axes (we do everything by command line), for example as in Table 8.

Once these command lines have been added, it is possible to plot the graph within the plotting function: "plot".

With these few guidelines, one should be able to plot anything necessary. However, there can be some subtleties depending on what is needed to be done.

Be careful that in a ternary system, it is of course necessary to add a condition concerning the third element in the initial calculation (for example x (C)), and C is added to the binary A-B.

Binary diagram

Generally, a starting point must be chosen, and it is mandatory to stand in a two-phase field (in fact it is necessary to find a tie-line) even if the current versions of ThermoCalc are much more reliable than the first ones).

Binary system

Calculations must be extended in the two directions: x (A) on axis 1 and T on axis 2. The "map" command is doing that and generally it is presented (in the post) as "emf A" on x-axis and "T" or "TC" on the y-axis.

Sometimes, it is useful to define several starting points; in such a case, the software calculates the whole diagram more easily (in general a single-phase domain cuts the phase diagram in two, and the software stops looking for other multiphase domains). Look at the appendices for that.

Enthalpy of mixing

The starting point must be carefully chosen (among other things, the temperature since it will remain constant for this calculation).

Also, it is mandatory to define the reference states well (generally, the liquids for the two phases at the temperature of calculation).

It is necessary to extend the calculation in one direction: x(A) on axis 1. We extend the calculation with a "step", and we present (in the post) "mf A" on the x-axis and "HMR (LIQUID)" on the y-axis.

Isothermal section

The starting point must be carefully chosen (among other things, the temperature since it will remain constant for this calculation). The same remark is for the multiple starting points (see appendices).

The calculation must be extended in two directions: x(A) on axis 1 and x(B) on axis 2. We extend the calculation with a "map", and we present (in the post) "mf A" on the x-axis and "mf B" on the y-axis.

The command "s-d-t Y" (set-diagram-type) (in the post) must be added in order to have a representation in the form of a triangle. It can also be useful to trace the "s-tie 1" ctie-lines (1 traces all the ctie-lines, 2 half, 3 a third, etc.). The name of the elements with the "s-cor" can be displayed by command as written previously, and use also lr and ll (for low-right and low-left) AND top (for the third element).

Isopleth section

Basically, it will be written in the same way as for a binary diagram. The calculation must be extended in two directions: x(A) on axis 1 and T on axis 2. The calculation is extended with a "map" and generally presented (in the post) "mf A" on x-axis and "T" or "T-C" on the y-axis.

On the other hand, the choice of the starting point is more complex and more sensitive. In the ternary U–Si–B system, the plot of isopleth at 20% B is done in the following conditions defined as in Table 9a.

On the other hand, if the drawing of an isopleth section between UAl2 and Nb is needed, the condition on the U/Al ratio is used such as in Table 9b.

In the most complicated cases, S in drawing the isopleth section between UAl2 and Nb3Al; a condition on the U/Al ratio again is needed, but before that the calculation of equation of the line is mandatory:

Saving the plotted figures

After having plotted the figures, it is possible to save them and in such a case, always in the post, it is mandatory first to define the plot format with the command "set-plot-format" followed by a number representing the graphic mode (the ? allows to list the formats available). Note No. 22 (graphics mode by default). EMF formats done with 19. The generated file can then be inserted into an editor application such as PowerPoint for example and be modified by the "ungroup" function.

Ternary phases

Open ThermoCalc and switch to the Console Mode if necessary. Then the command lines will be typed. Before calculations, the first thing to do is to recover the data related to the pure elements. For this, we can use a TDB module built before.

The ternary phases can be added first as stoichiometric (AxByCz) starting with a declaration of those phases according to Table 10 and the following lines will be added. So far, it has been seen how to create and use a ternary TDB from an optimized thermodynamic database published in an article. This base is then completed by adding a ternary phase and by modeling its solubilities in the binaries.

It remains to estimate the parameters XXX (in the enthalpy of formation) and YYY (in the entropy of formation). In the case of unavailability of enthalpy of formation in the literature (which is very often the case), a value must be found that should be consistent with the phase diagram.

First start with a value for XXX (and at the same time, set initially YYY to 0). In such a case, the knowledge of the equilibria at "low temperature" is 298 K, or even a little more is needed. Being referred to a mol of matter (and not a mole of formula), the enthalpy of formation must be negative; generally, this value is in the range -20.000 to -50.000 (which corresponds to the classic order of magnitude for intermetallics). If the phase is not stable enough (fairly negative enthalpy), it will not appear (case on the left below). If, on the contrary, it is too stable, it risks being in equilibrium with wrong phases (case on the right below). Of course, a correct value (case of the medium) must represent the equilibria identified experimentally. There is not a single value, but an area of value. We can also estimate the value of XXX by knowing the formation enthalpies of the phases with which our phase is in equilibrium.

The calculation process is done as follows:

− Setting a first value of XXX in the TDB as explained (which is recorded);
− Calculation of the isothermal section;
− Consequently changing the value of XXX in the TDB, etc.

...

For a good calculation, the phases must be added one after the other. When the XXX of the first phase is adjusted, we move on to the second phase, etc. one by one. The order of the ternary phases to be optimized depends on the stability (from the most stable to the least stable).

Then starting to adjust YYY (- entropy). It is mandatory that this value MUST BE POSITIVE! If YYY is negative, the phase will be stabilized with temperature, which is not correct.

The entropic parameter influences greatly the phase stability with temperature, and in particular its decomposition temperature such as melting and allotropy. In order to check the temperature, it is advised to write a macro corresponding to an isopleth section containing the phase and test it.

We will therefore adjust YYY in order to match the decomposition temperature measured by ATD or calculated ab initio. Otherwise, we can at least choose a value that is consistent with the mode of formation of the phase (congruent, peritectic or peritectoid).

Of course, as the value of YYY is increased, the phase is destabilized at high temperature, and therefore the decomposition/melting temperature of the phase is decreasing.

Once the value of one phase has been refined one turns to the next phase. When all the parameters are set, different isothermal/isopleth sections are plotted in order to verify. Sometimes, adding parameters for a second phase changes the balances originally set for the first.

Table 1

s-c: set-conditions (define the conditions for a calculation).

c-e: compute-equilibrium (calculation of equilibrium with the previous conditions), very important to analyze in order to get good calculation conditions.

l-e: list-equilibrium (shows the results of the calculation: phases, compositions, quantity).

g or go: goto (go to a module).

b or ba or bac: back.

exit (exit ThermoCalc).

In general, ThermoCalc is not breakage-sensitive, but it is better to write the atoms (and vacancies) in capital letters (AL, SI, NB, VA, etc.).

The software contains several modules that allow performing different types of calculations.

The first thing that must be done concerning the creation of a file called TDB file consists in gathering all the data concerning the pure elements in their state of reference and the different forms of binary phases (or ternary) in the frame that they will then be used in phase diagram calculation procedure.

The second thing to do is to recover the data relating to the pure elements. For this, we use the TDB module.

On the other hand, when relevant data exist in the literature, it is necessary to take them into account, and this explains the need to have good data in the binaries and good modeling before going up to systems of order n + 1.

Open ThermoCalc and switch to the Console Mode if necessary.

This creation is made through the command lines as in Table 11.

One will have to use this procedure (the first 5 lines) each time a database is opened.

We will see that when writing the macros.

This produces an editable TDB file.

References concerning this chapter.

- ThermoCalc website (http://www.thermocalc.com/support/documentation/)
- CALPHAD (Calculation of Phase Diagrams): A Comprehensive Guide, Edited by N. Saunders and A. P. Miodownik (1998) ISBN: 978-0-08–042,129-2
- Computational Thermodynamics: the Calphad Method, by H. Lukas, S. G. Fries and B. Sundman (2007) ISBN; 978-0-52-186 811-2

Regarding phase modeling:
Ionic liquid (M. Hillert et al., Metall. Trans. 16A (1985) 261–266).
Solid solutions (M. Hillert, J. Alloys Compds 320 (2001) 161–176).
Quasichemical liquid (A. D. Pellton et al., Metal. Mater. Trans. 31B (2000) 651–659).
Taking magnetism into account (W. Xiong et al., Calphad 39 (2012) 11–20).
Introduction to the Thermocalc software.

References

1. I. Barin, *Thermochemical Data of Pure Substances*, 3rd edn. (VCH Verlagsgesellschaft mbH, 1995); I. Barin, O. Knacke, O. Kubaschewski, *Thermochemical Properties of Inorganic Substances* (Springer-Verlag Berlin Heidelberg Gmbh, 2013)
2. B. Fabrichnaya Olga, K. Saxena Surendra, R. Pascal, F. Westrum Edgar Jr., *Thermodynamic Data, Models, and Phase Diagrams in Multicomponent Oxide Systems: An Assessment for Materials and Planetary Scientists Based on Calorimetric, Volumetric and Phase Equilibrium Data* (Springer Berlin Heidelberg, Berlin, Heidelberg, 2014). ISBN 9783662105047. OCLC 851391370
3. Thermocalc website (http://www.thermocalc.com/support/documentation/)
4. Z.-K. Liu, Y. Wang, *Computational Thermodynamics of Materials* (Cambridge, 2016)
5. A. Dinsdale, SGTE data for pure elements. Calphad **15**, 317–425 (1991)
6. P. Atkins, J. de Paula, J. Keeler, *Atkins' Physical Chemistry: International Eleventh Edition* (Oxford, 2018)
7. NIST-JANAF Thermochemical Tables, NIST Standard Reference Database 13. *Last Update to Data Content* (1998)

Chapter 5
Ab Initio Calculations in the CALPHAD Methodology, the Quantic Simulations: Slight Description and Use in CALPHAD

5.1 Introduction. «An Introduction to Numerical Simulations in Order to Approach Specialized Works in This Field»

Numerical simulations have won a nice place in the physical sciences. It has become common to talk about them as «numerical experiences». De facto, this expression defines a field of physics related to a combination of experimental and theoretical approaches: necessarily experiments are used to validate the results of calculations. On the one hand, it involves observation, as finely as possible, of the behavior of matter at the desired scale with a good knowledge of the problems, then numerical simulations can support experiments (or even replace them). On the other hand, the comparison of the results obtained with available experiments makes it possible to validate the relevance of the theoretical approaches whose simulations are only numerical implementations. In materials science, this type of approach has been developed to understand and explore the behavior of matter at the atomic scale. We will try here to present the uses and limits of numerical quantum simulations in this field. The principle of such calculations is based on solving the Schrödinger equation by using the density functional theory (DFT). It is well known that solving the Schrödinger equation is impossible. But, by using some approximations concerning electronic structure, it is possible. The pending question will be: which kind of approximation should we use?

Quantum calculations, as they stands, pave the way for the design of quantum materials for future technologies.

5.2 Resolution of the Schrödinger Equation

Solving the Schrödinger equation in a complex multi-ion, multi-electron system is very difficult in the case of a solid. It is only accessible through computational

© The Author(s), under exclusive license to Springer Nature Switzerland AG 2024   75
J.-C. Tedenac, *Thermodynamics of Crystalline Materials*,
SpringerBriefs in Materials, https://doi.org/10.1007/978-3-030-99027-5_5

approaches. Different steps should be used. The first step concerns the decoupling of ionic and electronic dynamics because their characteristic time scales differ by several orders of magnitude. In such a case, the adiabatic approximation is used with profit. Another very important step was taken from the work of Walter Kohn et al. [1]. They developed a theory of the electron density function. It reduces the intractable complexity of multi-body electronic interactions to a one-electron equation. This is determined by the exchange–correlation functional which depends only on the electron density. Although the form of this functional which would make the reformulation of the many-electron Schrödinger equation exact is unknown, the approximate functionals have been found to be very effective in describing the properties of many materials.

However, even with the simplifications introduced by DFT theory, the development of highly efficient and precise algorithms is necessary to solve the Kohn–Sham equations in technologically relevant systems. It's not a simple problem.

To date, Gaussian software is the most important tool for performing quantum chemistry calculations on molecules. The equivalent of the GAUSSIAN package for solid-state and materials physics is under development.

The Vienna VASP ab initio simulation package developed by Georg Kresse et al. [2–6] is one of the many initiatives in this direction and is very successful.

Let's go to the roots of quantum mechanics with are supposed well known before this chapter. As written before, solving the Schrödinger equation is actually impossible except when using approximations. Nevertheless, the principle of such calculations is based on solving the Schrödinger equation by using the density functional theory (DFT). The question now is: What is DFT? DFT is based on the calculation of the electron density ρ (~ r). In this approach, energy is expressed as a function of the electron density. So, if $Eo[\rho]$ represents the monoelectronic energy, $J[\rho]$ is Coulomb's electrostatic energy and the potential $V_{xc}[\rho]$ is the exchange and correlation energy which is an unknown function. Therefore $V_{xc}[\rho]$ is needed to be understood and calculated. The first step concerns the choice of the DFT function and after that the choice of potential which will be used. The energy is expressed throughout this equation:

$$\rho(\vec{r}) = \int_1 \cdots \int_{N-1} |\psi(\vec{r}_1, \ldots, \vec{r}_N)|^2 d\vec{r}_1 \ldots d\vec{r}_{N-1} \qquad (5.1)$$

Energy is expressed as the equation which is a functional of the density:

$$E[\rho] = Eo[\rho] + J[\rho] + V_{xc}[\rho] \qquad (5.2)$$

where

$Eo[\rho]$ monoelectronic energy.
$J[\rho]$ electrostatic energy (Coulomb).
$V_{xc}[\rho]$ exchange and correlation energy which is unfortunately unknown.

The choice of the density functional.

Table 5.1 The different types of approximations used in the DFT procedure

Approximations	Nature
LDA	Local density approximation: $V_{xc}[\rho]\,(xo) = f\,[\rho(xo)]$
GGA	Generalized gradient approximation: $V_{xc}[\rho]\,(xo) = f\,[\rho(xo), \nabla(xo)\rho]$
Meta-GGA	$V_{xc}\,[\rho]\,(xo) - f\,[\rho\,(xo), \nabla\rho\,\rho, \Delta n\delta\mu]$
Hybrid-GGA	With a small exchange as Hartree–Fock

In DFT, the choice of the level of approximation is made by choosing firstly a functional.

Different types of approximations can be used. They are presented in Table 5.1.

By using the Block theorem, the wave functions adapted to translational symmetry are written.

~k characterize the translational symmetry.

The basis used to describe the wave function of the system. Two big families:

- Atomic orbitals (Gaussian type orbitals).
- plane waves (with k is the wave vector).

$$\psi_{\vec{k}}(\vec{r}) = \sum_{m=1}^{N_{maille}} \exp\left[i(\vec{k} + \vec{K}) \cdot \vec{r}_m\right] \qquad (5.3)$$

The base size is fixed by the maximum value of the energy E of a plane wave (it is named cutoff). This function is adapted to crystallographic problems. Only one parameter is now obtained: the cutoff energy:

$$\varepsilon = \hbar^2 \left|\vec{k} + \vec{K}\right|^2 / 2m_e \qquad (5.4)$$

As it approaches the nucleus, the wave function is rapidly oscillating.

Far from the core it is monotonous. It entails some difficulties in the description of the two regions simultaneously. The solution to this problem is a calculation of all electrons with a large base. Concerning the pseudo-potential close to the nucleus, the wave function ψ is replaced by a pseudo-function which is less oscillating, Fig. 1. This allows reducing the size of the base (the cutoff) and the calculation time.

The elementary cell is repeated periodically in all directions of space. This modeling is ideal for the representation of an infinite periodic system (a perfect crystal). It is well suited to the use of plane wave bases. But in the case of processing 2D, 1D or non-periodic systems, as well as handling stacking faults and partial occupations, it is not so good.

The structure of the calculation is organized as shown in Fig. 2.

Fig. 2 The different files organizing the calculation

INCAR		INCAR: Type of calculation and calculation parameters.
POSCAR		POSCAR: Initial positions of atoms in the structure
	INPUT	
POTCAR		POTCAR: Pseudo-potentials used (nature and form)
KPOINTS		KPOINTS: Grid of the k-points used in the calculation
↓		
VASP		
↓		
CHGCA CONTCAR DOSCAR EIGENVAL OSZICAR OUTCAR vasprun.xml WAVECA	**OUTPUT**	These different Output files characterizes the table of calculated values

5.3   Case Study: The System Ti–Sn (Published in «CALPHAD: Computer Coupling of Phase Diagrams and Thermochemistry» (5.3))

The application of this DFT theory will be now described by using a real system. Tin–Titanium. This system is very interesting as it lye two different elements (one is Transition Metal and the second is p element, or post-transitional), and it shows some intermetallic compounds with different stabilities inside the convex hull. The density functional theory (DFT) calculations have been carried out using the Vienna Ab initio Simulation Package (VASP) in order to obtain the band structure and the thermodynamic potentials. The calculations use potentials constructed by the projector-augmented waves (PAW) methods. This method (potential and linear augmented plane waves) allows performing density functional theory calculations with greater computational efficiency. The Muffin–Tin approximation is an approximation of the shape of the potential well in the crystal lattice. This is most commonly used in simulations of the electronic band structure in solids. The approximation was proposed by John C. Slater (Slater's approximation). The Augmented Plane Wave (APW) method is a method that uses the Muffin–Tin approximation. It is a shape approximation of the potential well in the region surrounding a local minimum of potential energy. The basic approximation lies in the potential; it assumes to have spherical symmetry in the region considered and constant in the interstitial region. The wave functions are constructed by matching the solutions of the Schrödinger equation in each sphere with plane wave solutions in the interstitial region, and the linear combinations of these wave functions are then determined by the variational method. Many modern methods of electronic structure use approximation. Among them are the APW method, the linear orbital muffin–tin method (LMTO) and various functional methods of Green.

For the exchange–correlation functional, the generalized gradient approximation (GGA) [7] is applied. In the case the problem for the PAW potentials concern the 3d and 4s orbitals and the semi-core 3p orbitals. There they are treated as valence orbitals for Ti (3p6 3d3 4s1 electronic configuration) while 4d, 5s and 5p orbitals are used as valence orbitals for Sn (4d10 5s2 5p2 electronic configuration). A plane-wave cutoff energy of 300 eV for both elements has been taken as it is often used in such systems.

The calculations are performed for the experimentally observed compounds at their ideal stoichiometry. The structures which are generally observed as stable in the systems rely upon the early transition metals with p-elements of columns IIIB, IVB and VB.

Concerning the stable intermetallic compounds of the system, the calculated zero-temperature lattice parameters agree very well with those obtained experimentally at ambient temperature (x). Moreover, for the stable phases possessing degrees of freedom specific to the unit cell, the results of ab initio calculations show a really good agreement when compared with data obtained by structural analysis of X-ray diffraction. The composition dependence of the enthalpies of formation is in the convex hull which shows a slight asymmetry, indicating a slight asymmetry of the electronic densities. In order to understand the phase stabilities of the Sn3Ti5 compound in the possible different forms, the electronic densities of the state of the D88–Sn3Ti5 compound have been computed in order to do the same calculations with other compounds having this formula. The curve shows the hybridization of Sn(5p) states with Ti(3d) states. It appears that the stability of the intermetallic compounds in the Ti–Sn system is due to this hybridization.

The calculated formation enthalpy of the hexagonal Sn5Ti6 compound is slightly less exothermic than the value obtained by direct reaction calorimetry.

The first-principle calculations of the enthalpies of formation of the SnTi inter-metallic compounds have been done by using the Vienna Ab initio Simulation Package (VASP) [2]. The purpose of this work is to improve the knowledge of the Sn–Ti alloys, and to complement the available experimental data. In particular, these calculations could give an answer concerning the ground state for the Sn5Ti6 compound. The dataset is being incorporated into the thermodynamic database containing the Sn–Ti binary system.

One can note that the calculations have also been performed at given stoichiometry with other structures than those observed experimentally to check the possibility of precipitation of metastable phases during alloy processing. Such calculations are also helpful to predict the possible stabilization of the structures by ternary additions to the Sn–Ti system. Moreover, in the further CALPHAD calculations of X–Sn–Ti ternary phase diagrams, it is obviously useful to have the values of the enthalpies of formation in the Sn–Ti system of phases which are stable in the two other binaries X–Ti and X–Sn.

The ab initio calculations of enthalpies of formation have been performed in the Sn–Ti system. For the Brillouin zone integration, the Methfessel–Paxton [4] technique with a modest smearing of the one-electron levels (0.1 eV) is used. Care was taken that for each structure a sufficient number of k points for the Brillouin zone integration was chosen. For all the hexagonal structures, a gamma centered k-point grid was used. For the other structures, a Monkhorst–Pack [5] grid was used. In each case, the number of k points has been chosen in such a way that the number of irreducible k points multiplied by the number of atoms in the cell is of the order of magnitude of 500. A very good control of the quality of the calculation is the comparison of the lattice constants with those measured. In this example, there is a convergence of the values of the lattice parameters and of the total energy. All calculations have been performed using the "accurate" setting within VASP to avoid wrap-around errors. With the chosen plane-wave cutoff and k-point sampling, the reported enthalpies of formation are estimated to be converged with a precision better than ±0.5 kJ/mol of atoms.

All the results presented in this example were obtained by employing the computational settings previously described. In order to verify, we have performed additional calculations with alternative settings, checked the overall accuracy of the reported results and tested the influence of a particular setting.

It appeared that the inclusion of the semi-core states (Sn-4d electrons and Ti-3p) lead to an increase in the enthalpies of formation (less negative values) of 0.5–1.1 kJ/mol of atoms depending on the composition; the major effect is due to the d-Sn semi-core electrons, by increasing to 360 eV. The modifications of the enthalpies of formation which are appreciable are those of SnTi, D019, 0.010 kJ/mol of atoms and h-Sn$_5$Ti$_6$+ 0.015 kJ/mol of atoms. Switching from GGA to LDA (local density approximation) has the effect of making the calculated enthalpies of formation more negative, by 5.5 kJ/mol of atoms for SnTi$_3$-D019 and by 7.5 kJ/mol of atoms for h-Sn$_5$Ti$_6$, for example. The lattice parameters obtained by the LDA are reduced compared to GGA. In fact, the GGA calculations give a significantly better agreement with the experimentally measured lattice parameters as compared to results derived by the LDA.

The enthalpy of formation, ΔfH(SnxTi$_1$ − x), at zero pressure is obtained from the minimum total energy of the compound (expressed per atom) by subtracting the composition-weighted minimum total energies of pure body-centered tetragonal (A5) Sn and pure hexagonal close-packed (A3) Ti according to the equation:

The Emin values are obtained after volume relaxation; this implies zero pressure.

5.4 Conclusion

The enthalpies of formation of intermetallic compounds of the Sn–Ti system have been obtained using an ab initio method. The predicted ground state structures are consistent with those known to be stable at low temperature. The calculations allow us to confirm the stabilities of some compounds; therefore the orthorhombic structure of the Sn$_5$Ti$_6$ compound is not a ground state at low temperature. The calculations have also been performed for several crystal structures which have been found stable in bordering systems of Sn–Ti. The ab initio calculations have shown that the enthalpies of formation of Sn$_2$Ti in the CuMg$_2$-type and NiMg$_2$-type structures are only 1 kJ above the ground state line. The hexagonal structure, hP22, displayed by the Sn$_5$Ti$_6$ compound, is found only in three other systems, say Pb–Ti, Ga–V and Ga–Ta, where it is a high-pressure form. The structure displayed by the Sn$_3$Ti$_2$ compound was until now never found in binary systems. This structure appears clearly stable. The ab initio calculations have also shown that the enthalpies of formation of Sn$_2$Ti in the CuMg$_2$-type and NiMg$_2$-type structures are only 1 kJ above the ground state line. Obviously, this approach is very complementary to experimental results, if they exist. In all cases, this computational approach is sufficiently robust to give relevant results (Figs. 5.1, 5.2, 5.3, 5.4, 5.5 and 5.6).

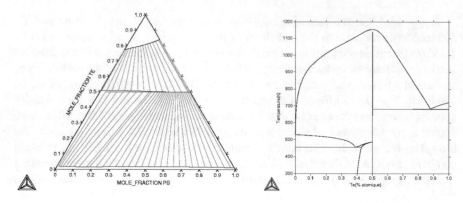

Fig. 5.1 .

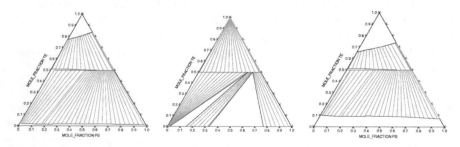

Fig. 5.2 .

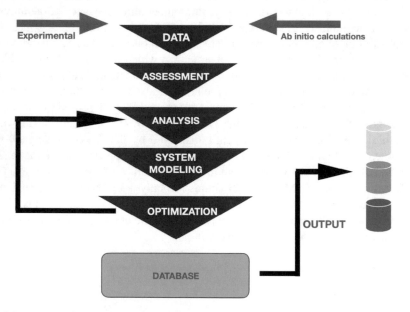

Fig. 5.3 .

Reaction type	equilibrium between liquid and three solid phases
eutectic [class 1]	liquid⇌α+β+γ
transitory reaction or 'peritectic' [class 2]	liquid+α⇌β+γ * this reaction is an intermediate between eutectic and peritectic. It is also sometimes named peritectic of class 2
peritectic [class 0]	liquid+α+β⇌γ

Fig. 5.4 .

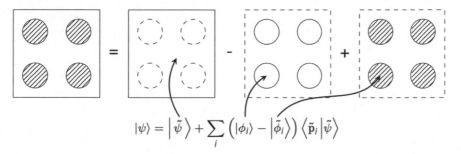

$$|\psi\rangle = |\tilde{\psi}\rangle + \sum_i \left(|\phi_i\rangle - |\tilde{\phi}_i\rangle\right)\langle\tilde{p}_i|\tilde{\psi}\rangle$$

Fig. 5.5 .

Fig. 5.6 .

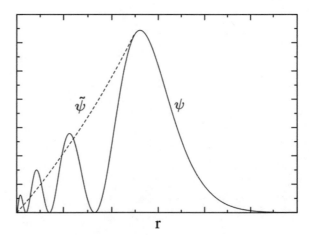

References

1. J. Willard Gibbs, Transactions of the connecticut academy of arts and sciences, *Equilibrium of Heterogeneous Substances* (1876)
2. P.E. Blöchl, Phys. Rev. B **50**, 17953 (1994)
3. G. Kresse, D. Joubert, Phys. Rev. B **59**, 1758 (1998)
4. J.P. Perdew, S. Burke, M. Ernzerhof, Phys. Rev. Lett. **77**, 3865 (1996)

5. M. Methfessel, A.T. Paxton, Phys. Rev. B **40**, 3616 (1989)
6. J.P. Perdew, Y. Wang, Phys. Rev. B. **45**, 13244 (1992)
7. U. Kattner, H.L. Lukas, G. Petsow, B. Gather, E. Irle, R. Blachnik, Z. fur MetallK. **79**, 32–40 (1988)

Conclusion

This book describes the present state in the determination and uses of the multicomponent phase diagrams in the scope of audiences who are not specialists in thermodynamics but wish to use it wisely in materials science. Obviously, complex phase equilibria calculations are presently very well developed in alloys and metallurgy and sometimes, but partially, in other material classes. In order to do this research, it is necessary to turn to these books and articles. As some pertinent tools were used during the last past years, it is very important to use them presently and the present book is a guide to choosing the best way to do it.

The knowledge of thermodynamic theory has been implemented in the last century. Thermodynamic modeling tools are very well known in the twenty-first century and can be supported by the creative use and behavior of all kinds of materials. The master idea is the current Gibbs energy minimization, and new advances were done in these areas.

New fields of applications need more multidisciplinary research in all new directions. Their knowledge will be necessary presently and in the same procedure: the CALPHAD procedure. It should be open.

Because of pregnancy of some problems, the CALPHAD method is clearly necessary.

Validation of multicomponent databases, implementation of the models and applications in the validation of experimental data are the most recognized directions of the present development.

This is the guideline of this book, and it explains the division into these five chapters. Each of them presents one of the facets of each part developed in the CALPHAD method: Crystallography, fundamental thermodynamics, phase equilibria, CALPHAD methodology and ab initio calculations.

© The Author(s), under exclusive license to Springer Nature Switzerland AG 2024   85
J.-C. Tedenac, *Thermodynamics of Crystalline Materials*,
SpringerBriefs in Materials, https://doi.org/10.1007/978-3-030-99027-5

Bibliography

1. *CALPHAD (Calculation of Phase Diagrams): A Comprehensive Guide*, ed. by N. Saunders, A.P. Miodownik (1998). ISBN: 978-0-08-042129-2
2. *Computational Thermodynamics: The Calphad Method*, ed. by H. Lukas, S.G. Fries, B. Sundman (2007). ISBN: 978-0-52-186811-2
3. M. Hillert et al., Metall. Trans. **16A**, 261–266 (1985)
4. M. Hillert, J. Alloys Compds **320**, 161–176 (2001)
5. A.D. Pelton et al., Metal. Mater. Trans. **31B**, 651–659 (2000)
6. W. Xiong et al., Calphad **39**, 11–20 (2012)
7. Introduction to the Thermocalc software. https://thermocalc.com, https://thermocalc.com
8. S. Qin et al., Thermodynamic modeling of the Ca–In and Ca–Sb systems supported with first-principles calculations. Calphad **48**, 35–42 (2015)
9. G. Kresse, J. Furthmüller, Comp. Mater. Sci. **6**, 15 (1996); Phys. Rev. B **54**, 11169 (1996)
10. J.P. Perdew, J.A. Chevary, S.H. Vosko, K.A. Jackson, M.R. Pederson, D.J. Singh, C. Fiolhais, Phys. Rev. B **46**, 6671 (1992)
11. C. Colinet, J.C. Tedenac, S.G. Fries, Computer coupling of phase diagrams and thermochemistry. *CALPAD* **33**, 250–259 (2009)

Printed in the United States
by Baker & Taylor Publisher Services